Errors in Evidence-Based Decision Making

Errors in Evidence-Based Decision Making

Improving and Applying Research Literacy

Robert W. Janke and
Bruce S. Cooper

ROWMAN & LITTLEFIELD
Lanham • Boulder • New York • London

Published by Rowman & Littlefield
A wholly owned subsidiary of The Rowman & Littlefield Publishing Group, Inc.
4501 Forbes Boulevard, Suite 200, Lanham, Maryland 20706
www.rowman.com

16 Carlisle Street, London W1D 3BT, United Kingdom

British Library Cataloguing in Publication Information Available

Library of Congress Cataloging-in-Publication Data

Janke, Robert W.
Errors in evidence-based decision making : improving and applying research literacy / Robert W. Janke and Bruce S. Cooper.
pages cm
Includes index.
ISBN 978-1-4758-1080-6 (cloth : alk. paper)—ISBN 978-1-4758-1081-3 (pbk. : alk. paper)— ISBN 978-1-4758-1082-0 (electronic)
1. Research—Methodology—Textbooks. 2. Decision-making—Textbooks. 3. Science—Methodology—Textbooks. 4. Errors, Scientific—Textbooks. I. Cooper, Bruce S. II. Title.
Q180.55.M4J36 2014
001.4'2—dc23
2014007516

∞™ The paper used in this publication meets the minimum requirements of American National Standard for Information Sciences Permanence of Paper for Printed Library Materials, ANSI/NISO Z39.48-1992.

Printed in the United States of America

Contents

Preface

The whole of science is nothing more than a refinement of everyday thinking.
—Albert Einstein

Whenever I look out my window on a "damp, drizzly November" day, I think of Ishmael in *Moby-Dick*, who escaped to the sea to cure his blues. Unlike Ishmael, I get the blues, as I feel adrift in the sea of dreary textbooks used in teaching introductory research procedures to students over the past 30 years.

These introductory textbooks are promoted with a glowing list of adjectives, including unique, impressive, practical, readable, and on and on. Some of these textbooks are so popular that there have been more than 10 editions. Unfortunately, these textbooks overwhelm the student with a mind-numbing catalogue of research procedures, which are promptly forgotten after completion of the textbook.

Any one of these textbooks could be titled *Sleepwalking through Research*. To paraphrase the comments at Caesar's funeral, a good introductory textbook should come to educate students, not to bury them. The important issue is not how much information textbooks throw at students, but how much information students retain and apply. Our students have expressed the following multiple concerns that inhibit their ability to become fully engaged in learning about research procedures.

First, students often fail to see the personal importance of WHY they need to learn about research procedures. A simple response, "Because this will be on the test," just does not get students engaged. Human nature makes us more sensitive to perceived loss than to gain. Therefore, a research textbook that provides examples of the consequences of erroneous use of research procedures can provide an epiphany to increase student engagement in learning about appropriate research procedures and avoiding errors in gathering and interpreting evidence.

Second, students have anxiety over learning about research procedures. A textbook that uses simple and familiar language with some humor can reduce this anxiety. Humor should not overwhelm or distract, but rather highlight the flavor and tenderize the topic. Research procedures is a meaty topic, but it must be more easily digestible to students!

Third, students do not see the generalizability of research procedures to multiple situations outside of the classroom. Research procedures are not the exclusive province of one academic field. Everyday life is multidimensional, involving multiple situations with multiple types of evidence and multiple types of decisions. A textbook that includes multiple and diverse real-world examples can increase engagement by showing the generalizability of research procedures to the global society beyond the classroom.

Fourth, students often feel confused because textbooks present information perceived as a series of separate and unconnected topics. Cognitive and pedagogical sciences indicate that student engagement is enhanced when scaffolded with clear learning objectives, development of prerequisite skills and dispositions, and descriptions of the sequence and interrelatedness of the topics. A textbook that incorporates these additional suggestions can enhance students' engagement and learning.

Fifth, fatigue and withdrawal can result when introductory textbooks have 600 or more pages and provide too much detailed information (TMI) that overwhelms students. An introductory textbook should provide information about the breadth of important procedures that provide the scaffolding and foundation for later, more detailed and comprehensive study. A shorter introductory research textbook may be more appropriate to engage students and develop the prerequisite dispositions for further study.

Sixth, the mathematics in most introductory textbooks often intimidates students. An introductory textbook that avoids detailed mathematical material and uses simple language will make students feel less threatened and more likely to engage in learning about research procedures and in identifying and reducing erroneous evidence.

Where do we find an introductory research textbook to address the above concerns? After years of searching for such a book, we began to appreciate the frustration of explorers searching for the Fountain of Youth. The textbook that we wanted did not yet exist. We then started to come to a rather frightening conclusion, suggested by the many voices that we hear (relax; we are getting help for this problem). We should write a textbook that addresses the concerns of our students. Wow! What a challenge. Even Meatloaf was satisfied with much less when he sang, "Two out of three ain't bad."

If we can address those concerns, then we will have done, as in *A Tale of Two Cities*, "a far, far better thing" than we have ever done before. We just hope that we did not stick our necks out too far in trying to write a textbook. (Where do we get all of these bad jokes?)

After deciding to write a textbook, we bid farewell to our family and friends and said that we would look forward to seeing them again in a few years, when the textbook was finished. Once the cheers at not seeing us for a few years had subsided, we gathered our materials and started to write.

We wandered through the literature and available textbooks in search of ideas and directions to achieve the goal. In our search, we tried to follow the practical advice of wise persons such as Yogi Berra, who said, "When you come to a fork in the road, take it," as well as the poetic advice of Robert Frost that we should take the road "less traveled by." Unfortunately, we forgot our academic GPS at home, and we took the wrong road at the fork. In writing this textbook, we had to abandon many of the ideas along the way.

After years of thought and listening to the thousands of voices that we hear (we forget, did we say that we are getting help for this problem?), the book is now complete. We always dreamed of writing the Great American Textbook. What is the spelling of that famous prize for literature: Nobel or Noble?

We hope that this introductory research book will engage students and their teachers by showing that research is a dynamic process, integrated and interconnected with the lives of every person past, present, and future. We further hope that this book will motivate students by presenting research as a life force that has contributed to making us who we are and, importantly, that can contribute to making us who we will become.

Acknowledgments

Robert Janke

Drs. Michael Dwyer and Charles Levin provided encouragement and constructive suggestions for writing this book. Dr. Kenneth Schneck made a significant contribution by introducing me to the co-author, Dr. Bruce S. Cooper. Dr. Cooper's energy and passion for writing are remarkable. Pastor David Collings inspired dedication to completing this book.

My wife, Kathy, gave me support and understanding during the many months of work involved. My late mother, Elsie, gave so much comfort to me and took great joy in my efforts. My regret is that she is not here to see the finished product.

This book is dedicated to my sons, Robert, Ryan, Kevin, and Daniel; my daughter-in-law, Emily; and my granddaughters, Anna and Juliet. My hope for all of you is that your lives will be free from error.

Introduction

This introductory book provides a nonmathematical and succinct description of research procedures as a foundation for more comprehensive presentation in later textbooks or courses. Each of the chapters in this textbook presents a different, but interrelated, issue and research procedure. These issues have not adequately been presented in previous textbooks and include

- the necessary dispositions of researchers,
- a comprehensive listing of common errors in research,
- examples of errors from multiple fields of study and from everyday life,
- the identification of misinformation, and
- the identification of personal bias in interpreting information.

The writing style of this introductory text follows suggestions both from cognitive sciences and from our students over the past 30 years to enhance learning and retention. The writing style emphasizes succinct descriptions and sentences and paragraphs of shorter length to sustain attention and facilitate learning with smaller "chunks" of information. The attempt is to present the information in a language and style that will be easier for students to understand. We will leave the complex and jargon-filled language for other textbooks.

We include a complete listing of hundreds of types of errors that occur in the application of different research procedures. In the interest of keeping this textbook to a succinct introduction to research, only representative examples of errors within each procedure are included. Unfortunately, we experience countless examples of errors in research every day.

The organization of this text also incorporates suggestions from the cognitive sciences and from students, in that each chapter follows this sequence:

- Significance—Why is this topic important?
- Learning Objectives—What do I need to know?
- Prerequisites—What should I already know?
- Concepts—Can I get a description of what I need to understand?
- Common Errors —What types of errors?
- Examples—Can I see some examples of the consequences of errors?
- Reducing Errors—How can I reduce errors?
- Best Practices—What are the best ways to reduce errors?
- Interrelation to Other Topics—What else do I need to know?

NOTE TO STUDENTS

Examples from multiple academic fields enlighten rather than inflame. The examples are poignant, with many resulting in the loss of life. The examples should provide opportunities for critical thinking in separating fact from inference and understanding the consequences of errors in the use of research procedures. The goal is to help us make the history that we will have better than the history we are now living.

WARNING LABEL: Do not attempt to read this textbook if you suffer from any of the following conditions:

Narrow mindedness, know-it-all-ism, difficulty separating fact from fiction, inability to change your beliefs when the facts change, allergy to uncertainty, low threshold of pain in having to think, addiction to easy answers and half-baked solutions. Also do not attempt to read if your favorite pet is a dogma.

DISCLAIMER: No humans or animals were hurt in the writing of this book, just a few sacred cows.

A RESEARCH EXAMPLE FROM LONG AGO

Example is the school of mankind, and they will learn at no other.

—Edmund Burke

Few things are harder to put up with than the annoyance of a good example.

—Mark Twain

The sounds of their feet shuffling on the parched earth were as grating as the sound of sandpaper rubbing against a rock. The pangs of hunger throbbed like an open wound. Neither the blisters on their feet nor the sharp hunger pangs could stop these three early humans as they wearily continued the search for food to survive in the desolate landscape.

Obit, Link, and Ugh continued their search for food until they saw nothing else but a lonely bush with small, round red berries and larger blue berries. The three had never before seen a bush like this. They were not sure what to do. Were the berries safe to eat? There was a risk, but the hunger was so great that Obit decided he had to know. He took the chance and ate some red berries. Within moments, he was writhing in pain and agony; soon, he was dead.

Link and Ugh were frightened. What should they do? The only other thing they saw to eat was the large blue berries on the bush. If the small red berries were not safe to eat, then would the large blue ones be safe? Why were there two different sizes and colors of berries? How could they know whether the blue berries were safe, seeing that no one they knew had any experience with or knowledge about the berries?

Link looked at the bush again and, in desperation, ate one of the larger blue berries. It tasted good, and he no longer felt hungry. He did not understand why there would be two different sizes and two different colors of berries on the same bush. Was it the size or the color of the berry that was important? They looked and looked, but they could not find any large red berries or small blue berries. They did know that large blue berries were safe to eat, but small red berries were not.

They could not understand why it was so, but the results of their search showed that only the large blue berries were safe. They gathered all of the large blue berries and hurried back to the small village to report what they had found. On the way, Link got lost; he was never found.

When he returned to the village, Ugh described the search, and how Obit had eaten the red berries and Link had eaten the blue. The villagers held a brief ceremony, which they called Obit the Berry, to recognize Obit's life. The ancient rite to honor a deceased person is now known as the "obituary."

Link's contribution in finding food to save humanity is important. To this day, people still are trying to find the missing Link.

The story told by Ugh is also important, and Ugh is celebrated. Any story told with warped humor causes this response: What a story, Ugh!

This almost-true story of Obit, Link, and Ugh illustrates that research procedures throughout human history increase knowledge about the world around us. For our early ancestors, obtaining accurate evidence and avoiding erroneous interpretation of the evidence were literally matters of life or death.

We are all descendants of ancestors who successfully used research procedures and accurate evidence. Modern life does not always present us with situations involving life-or-death decisions, but our decisions certainly affect the quality of our life and societal well-being. Today, research procedures are more formally understood as the activities comprising the Scientific Method. This method has immense power and is a fundamental way to gain knowledge of the world around us. But then why have we not always had success using this method, either in our personal lives or as a society? The challenge is to avoid errors in the application of these procedures in making real-world decisions.

Chapter One

Identifying Errors

He who would distinguish the true from the false must have an adequate idea of what is true and false.

—Benedict de Spinoza

Nothing is more damaging to a new truth than an old error.

—Johann Wolfgang von Goethe

Man errs as long as he strives.

—Johann Wolfgang von Goethe

People can be wrong in so many ways.

—Ben Goldacre

The war was started as the result of a mistaken intuitive "calculation."

—Hasegawa Nyozekan

When I make a mistake, it's a beaut!

—Fiorello LaGuardia

The method of scientific investigation is nothing but the expression of the necessary mode of working of the human mind.

—Thomas Huxley

I research; therefore I am.

—Anonymous (OK, you probably guessed it was us)

Science is built up with facts.

—Jules Henri Poincaré

Facts are stupid things.

—Ronald Reagan

Everyone is entitled to their own opinions, but not their own facts.

—Senator Daniel Patrick Moynihan

SIGNIFICANCE

Every person, business, or society has made errors in the form of inaccurate beliefs or decisions. These errors can occur at any stage of the processes that we use to gather information or evidence. They can reflect inaccurate or incomplete (weak) application of research processes or interpretation of research results. The errors can also be the result of our own disposition or resistance to information and even our errors in thinking or interpreting the information.

Identifying and then reducing these errors can improve both the quality and the quantity of results. These improvements can include reductions in time, cost, effort, conflict, disease, and so many other possibilities. Just think of the money saved by not having to put erasers on pencils!

Realistically, it may not be possible to eliminate all errors. Reducing the frequency or the severity of errors is an attainable goal and remains a primary focus of this textbook.

The Scientific Method is a process that has multiple embedded procedures to help reduce errors in gathering and interpreting information. Appropriate application of these procedures may offer the single best opportunity to reduce errors. Inaccurate or inappropriate application of these procedures may create errors.

This book presents important concepts involving appropriate procedures within the Scientific Method. The assumption is that we first need to understand what is appropriate to identify errors. Examples of inappropriate and erroneous applications of these procedures are included to reinforce the importance of reducing errors.

LEARNING OBJECTIVES

1. Understand that an error is a wrong belief or decision that results from inaccurate or incomplete (weak) application of research procedures or interpretation of research results.
2. Understand that errors are at the individual, group, or societal level.
3. Understand that errors may have immediate or longer-term consequences.

4. Understand that there are personal and societal consequences of using erroneous evidence.
5. Understand that errors may be intentional or unintentional.
6. Understand that identifying errors is a continuous process of monitoring and modifying beliefs and decisions.
7. Understand that reducing errors improves the quality of the evidence that we use to make decisions.
8. Understand that there are legal requirements to avoid erroneous behaviors and decisions.
9. Understand that there are moral admonitions to understand and to avoid errors.

    a. "Who can understand his errors?" (Psalms 19:12).
    b. "So whoever knows the right thing to do and fails to do it, for him it is sin" (James 4:17).

10. Understand that we may be more receptive to learning about appropriate research procedures if we see everyday examples of erroneous evidence resulting from the misapplication or misinterpretation of research procedures.
11. Identify the major steps in the Scientific Method.
12. Understand that research procedures are the activities conducted and embedded within the Scientific Method.
13. Understand that the research procedures of the Scientific Method are applicable to everyday decisions.
14. Understand that the structure of most journal articles reflects the major steps of the Scientific Method.
15. Understand that research provides tentative rather than absolute conclusions.
16. Understand that research seeks to obtain verifiable facts and evidence.

PREREQUISITES

1. Accept that there may be errors in any evidence.
2. Have a willingness to identify errors in evidence.
3. Accept that research procedures attempt to obtain verifiable facts and evidence.

CONCEPTS

All of us want to make the best choices and decisions. Unfortunately, we sometimes make erroneous or incorrect interpretations or decisions and then wonder how we could have made such an error.

What errors have we made? How serious was the error? Did we use inaccurate evidence or ignore accurate evidence? How long did it take to recognize that we made an error? Are we afraid to admit an error? Do we try to correct the error? Did we blame others rather than ourselves?

Each of us has made errors in our personal lives. We have made errors in selecting a career, a significant other, or even which car to purchase. Each of these errors has a financial, professional, or interpersonal consequence that we want to avoid. We will see in this textbook that correct application of research procedures can reduce these errors.

Much of the available information—and even many research studies—has errors that can contribute to erroneous beliefs and decisions. Some common errors involve the use of inaccurate data-gathering procedures, biased selection of the participants in a study, and lack of control for other potential explanations.

Errors as defined by this textbook are wrong beliefs or decisions that result from inaccurate or incomplete (weak) applications or interpretations of research procedures or research findings. The errors can be intentional (fraud, malfeasance, personal bias) or unintentional (ignorance or forgetfulness). The consequences of errors can be immediate or longer term and adversely affect individuals, groups, or society.

Regardless of the academic field or personal issues, the importance of reducing errors is obvious. Errors are an inevitable part of everyday life, and they can be financially costly and have adverse effects upon the quality of lives.

Our Founding Fathers' faith in a representative form of government assumed that well-informed citizens would use reason in civic participation. The First Amendment for freedom of the press assumes an accurate base of knowledge for decisions. However, freedom of the press does not guarantee the avoidance of errors and falsehoods. How much of the evidence that we receive on a daily basis is free from error? How can we recognize these errors?

Like zombies in horror movies, errors walk among us. We make errors every day. Have we taken any wrong turns driving today? Did we get into the slow-moving checkout line at the store? Did we say something that we regret? Did we make an error balancing our checkbook? Did we miss some questions on our last examination? Did these errors have any significant consequences for us?

Errors made by others can affect us. Many errors are made in individual credit reports. Have we checked to see if any errors exist in our reports? Could denial of our loan application be the consequence of an erroneous credit report? Errors can be relatively minor, but some can make the difference between life and death. Friendly fire can lead to a soldier being killed by his or her own troops.

What is the rate of errors in things such as medical diagnosis, mail deliveries, or marriages? Does it make a difference if the error affects others, but not us? Did the letter or check really get lost in the mail, or is this just a convenient excuse? Over 40,000 deaths in the United States each year occur because of traffic accidents. The shuttle *Challenger* exploded because of design errors in the quality of a component part.

The inevitability of errors means that each of us has to be willing to find errors and then modify beliefs or decisions to minimize them. A football team changes strategy at half time, after identifying the errors made in the first half. Did the team benefit from identifying and correcting or preventing errors in the second half? Did they win?

Legal requirements can help to avoid errors in documents submitted by individuals and companies. We often hear mentions in television courtroom dramas of telling "the truth, the whole truth, and nothing but the truth." Have errors been made in courtroom testimony or evidence? Do we think that these errors would affect the decisions of the jury?

Moral and biblical admonitions require understanding and avoiding errors. Correctly following God's guidance is doing the right thing. How many of us are without error in this regard? Are we selective in which admonitions we follow? Could we be in error in our interpretation? Could religious leaders make erroneous interpretations of evidence?

What are ways to reduce the chances of error? Correct application of the different research procedures presented in this textbook can reduce the probability of making errors. Research provides only answers that are tentative and not absolute, so we must constantly be vigilant to the possibility of errors in any decision that we make. This book provides hundreds of specific types of errors that can occur.

Understanding the research procedures or activities in the Scientific Method is an important skill and is included in the curriculum of many schools. For example, a recent booklet from the Ohio Department of Education, *A Standards Guide for Families—What Is Expected in Grade K*, states that kindergarten students should know how to gather information and use data analysis and probability to interpret results.

OMG: we now might have a tough time passing kindergarten! Can we still fit into the chairs? At our age, we are glad that kindergarten still includes naps.

The unanswered problem is why so few of us apply research procedures in everyday life. Errors can happen in any step of the research process. This textbook addresses this issue by helping us to find the errors or what can go wrong, providing examples of the consequences of errors and suggestions for best practice to prevent errors.

The Scientific Method is around us in many facets of our lives. For example, the major sections of articles that we see today in most scholarly journals still follow the sequence of the Scientific Method and can be illustrated (as shown in parentheses below) from the story of Obit, Link, and Ugh.

The Scientific Method

a. *Introduction to a journal article*

1. Present a need in which there is a discrepancy between the current and a desired status (hunger).
2. Present a problem: how to meet that need (we need to get food).
3. Ask a question: What are ways to solve the problem (where can we find safe food to eat)?
4. Determine whether the experiences of others can answer the question (did any other person eat this food?).
5. If the experience of others cannot answer the question, make a hypothesis or a prediction as to what we expect to find (red berries are safe to eat).

b. *Methods section of a journal article*

6. Gather information (collect the red berries).
7. Test the prediction (eat a red berry).

c. *Results section of a journal article*

8. Analyze data and determine support or refutation of the prediction (Obit dies after eating the red berry).

d. *Discussion section of a journal article*

9. A conclusion is made for future decisions (do not eat the red berries).
10. If the need is not resolved, repeat the process to determine any other possibilities or if any errors were made (collect and eat the blue berries).

The procedures or steps of conducting research indicated by the Scientific Method are in the fabric of our lives; we just didn't know it. Did we ever watch a baby conduct research by using different senses to learn about an object? Babies place many objects into the mouth to taste. Were some objects tasty, but others were not? That becomes a personal and unpleasant research experience. Are we getting nervous about this textbook and want to gnaw on it to learn more about research? Be our guest—just let us know how it tastes. By the way, a dry red wine goes best with page 14.

Do we ever wonder about some of our errors in decisions and say, "What was I thinking?" This book will help us to identify errors so that we will not have the recurring regret of saying to ourselves, "If I only knew then what we know now."

If you already correctly use the research procedures, congratulations. If we could afford to do so, we would get commemorative T-shirts. We would give others the shirts off our back, but we're not certain tie-dyed shirts are back in fashion.

All of us have experienced and will experience the research procedures many times, as either a subject of research or a researcher. Here are some ways in which we have been the subjects of research in different academic fields or commercial interests (as shown in parentheses):

- an advertisement that says "sale" (marketing)
- talk-radio shows that use emotionally labeling words such as "socialist," "values," "conservative," or "liberal" to generate controversy and gain market share (communication, political science)
- another person who says, "If you really love me you would" (interpersonal relations)
- funeral directors who create feelings of guilt by asking, "Don't you want the best casket for your parent?" (psychology)
- religious leaders who manipulate us with their interpretation of parts of the Bible (religion)
- our study habits as students based on what we thought the teacher or the professor would include on a test (education)
- what our peer group has established as standards for dress and behavior (interpersonal relations)

Here are some examples of topics for which we can conduct research within an academic field (as shown in parentheses):

- instructional strategies to improve student achievement (education)
- marketing strategies to increase sales (business and marketing)
- campaign slogans to gain an increased number of votes (political science)
- interventions to increase appropriate behavior and adjustment in clients (psychology)
- ways to rehabilitate criminals (sociology and criminal justice)
- ways to organize and deliver sermons to engage worshippers (religion)
- ways to diet and exercise (nutrition, health, and physical education)

- ways to purposefully inflame and create controversy to gain market share in talk radio and cable TV (communication)
- ways to influence our parents to get what we wanted as adolescents (how did real life sneak into this list?)
- ways to maintain a relationship with a significant other (interpersonal relations)

The Scientific Method is one way to obtain knowledge that can help us to understand the past and present to make better future decisions. The method provides an opportunity to obtain verifiable information that can help us to identify when we are wrong and to identify causes or patterns that we have not yet noticed.

Erroneous research creates confusion and leads to inaccurate decisions that may have significant adverse effects in our lives and in our society. Identifying common errors in each component procedure of the research process can help each of us.

The application of research procedures provides only a tentative conclusion. Researchers cannot control all possible influences upon an event, so it is incorrect to state that research "proves" something.

The possibilities for obtaining answers from research are as varied and changing as the turning of a kaleidoscope. Research is a continuing process of exploration, and researchers must be willing to accept results on a tentative basis. We cannot be entirely certain of the results from any research, and we must accept the conclusions in terms of probability rather than certainty. We have to accept, with humility, the research procedures of the Scientific Method as one way given by God to explore the mysteries of life.

COMMON ERRORS

1. Using intentionally deceptive, fabricated, falsified, or misleading information or malfeasance in application of research procedures (contributes to errors in interpreting research results)
2. Unintentional bias, ignorance, or forgetfulness (contributes to errors in application of research procedures and interpreting research results)
3. Using inaccurate or incomplete information (contributes to unintentional inaccurate application of research procedures and interpretation of research results)
4. Failure to acknowledge the error and continuance of the erroneous beliefs and decisions
5. Ignoring consequences of errors at the personal, group, or societal level
6. Ignoring long-term adverse effects

7. Not applying continuous monitoring for errors
8. Omitting some steps in the sequence of procedures of the Scientific Method in collecting data and making decisions
9. Ignoring all or selected facts in decision making
10. Inaccurately concluding that research "proves" or "disproves" an event

EXAMPLES

- *The New York Times* reported that even in 2013, 50% of Americans believed or were not sure that Saddam Hussein was involved in the September 11, 2001, attack on America. (Amazing how long an erroneous idea can last.)
- John Ioannidis, in an article in the *Journal of the American Medical Association*, reported that between a third and half of the most acclaimed peer-reviewed research articles in medicine are untrustworthy and not replicated because of errors made in the original research. (We get sick just thinking of all of the people who got ill from poor practice.)
- As reported in *The New York Times*, Phillip Sarrel and David Katz, Yale University professors, statistically analyzed mortality data and projected that nearly 50,000 women who had a hysterectomy in their 50s and 60s died because because they avoided hormone replacement therapy after the surgery. Resistance to taking the therapy was an erroneous conclusion from an incomplete earlier study funded by the Women's Health Initiative.
- Daniel Kahneman reported, in *Thinking, Fast and Slow*, the results of a study that indicated physicians who were "completely certain" of their diagnoses for patients in the intensive care units were wrong 40% of the time based upon autopsies. (There was reluctance to indicate how many patients died because of errors in diagnosis. We wonder if they just tried to bury the evidence?)
- The Bureau of the Census has indicated approximately 47% of all first marriages end in divorce. What are the consequences of these errors in first marriages?
- *The Plain Dealer* reported that Yankelovich Partners found 37% of Americans believe in astrology because we all look for cause-effect relationships even when they do not exist and then rationalize when they do not work. Are there any verifiable facts?
- Nicholas Kristof reported, in *The New York Times*, that about 50% of Americans think that humans lived at the same time as dinosaurs. (Actually, we are so old that one of us can remember having a dinosaur for a pet. His name was Dino.)

- *The Plain Dealer* reported that Jon Miller, a Michigan State University professor, presented findings that only about 10% of Americans know enough to understand newspaper reports about science, and that only about 40% of Americans accept the theory of evolution. That percentage signifies the second-lowest acceptance of the theory of evolution in developed countries in the world.
- Public Policy Polling found that even in 2013, 28% of Americans believed or were not sure that Bigfoot or Sasquatch exists. (You would believe if you came to one of our family reunions.)
- The Internal Revenue Service indicates that hundreds of thousands of errors occur in IRS tax returns each year. Are these intentional or unintentional errors? (Are we good at fuzzy math?)
- Jonah Goldberg reported, in *USA Today*, that about 50% of Americans believe that Karl Marx's principle of "from each according to his ability, to each according to his need" is in the Constitution of the United States. (Would a "strict constitutionalist" also be a communist?)

Still don't believe us that erroneous use of research procedures contributes to incorrect decisions? Here is a real-world decision that, unfortunately, is made many times each day by adolescents in this country and illustrates many of the research procedures that need to be considered:

Adolescents may have a need to explore or engage in sexual relations. The problem is how to prevent a pregnancy. This example illustrates errors in different research procedures.

The question is from whence to gather information: from friends? Did the friends know anything? Did the adolescents get the information from a good sample of friends? Was there accurate information or were there inaccuracies, such as "We cannot get pregnant if we have sex while we are standing"?

Did the adolescents then form a hypothesis and predict that they could not become pregnant the first time they had sexual intercourse? Did the adolescents drink alcohol? Did the variable of alcohol mediate and affect the ability to make an accurate decision? Did the adolescents test the prediction and have sex while standing? What were the results of having sex? Was the decision correct or incorrect? Was the pregnancy an example of an error caused by using inaccurate research information? Do we see many other errors made by the adolescents in applying research procedures? Was there another outcome or dependent variable, such as a sexually transmitted disease?

REDUCING ERRORS

As we can see, many types of errors can happen in different situations. Some of these errors may appear to be minor and even amusing, while other errors can literally mean the difference between life and death.

Regardless of the potential severity of the errors, reducing errors is an important task for each of us. If reduced-fat foods are better for us, then is reducing errors better for decision making?

BEST PRACTICES

- Follow the procedural steps of the Scientific Method.
- Examine each step of the research process for different types of errors.
- Continually monitor for errors in each step of the research process.
- Recognize the possibility of different types of errors.
- Consider research results as tentative rather than absolute.

The Scientific Method provides appropriate procedures for reducing errors. The chapters of this textbook describe those appropriate procedures and provide examples of inappropriate or erroneous application of those procedures.

INTERRELATION TO OTHER TOPICS

We have learned a definition of an error and that errors are all around us. We know that errors are inevitable and that they have consequences. How do we prevent these errors? The Scientific Method provides an opportunity to collect accurate evidence for making decisions.

Knowing the appropriate procedures of the Scientific Method described in this book is necessary but not sufficient. We still can make errors, and a significant reason is our own disposition or attitude when confronted with verifiable factual evidence that is contrary to our existing beliefs or behaviors. Some of us ignore or dismiss the factual evidence, while some of us accommodate and change our beliefs and behaviors.

Existing research textbooks give scant attention to the necessary dispositions for gathering and interpreting research evidence. The next chapter presents some needed dispositions for preventing errors and those dispositions that contribute to erroneous beliefs and behaviors.

Chapter Two

Dispositions and Ethics

"Know Thyself: Do We Cheat because We Can?"

When the facts change, I change my mind: What do you do?

—John Maynard Keynes

Minds are like parachutes. They only function when they are open.

—Sir James Dewar

The greater the ignorance the greater the dogmatism.

—Sir William Osler

True science teaches, above all, to doubt and be ignorant.

—Miguel de Unamuno

The best lack all conviction, while the worst are full of passionate intensity.

—W. B. Yeats

Most ignorant of what he's most assur'd.

—Shakespeare

It is good to love the unknown.

—Charles Lamb

What is the quality of a belief that exists simply because it's enforced?

—Maureen Dowd

If we cannot admit error, feel remorse and address problems, what future can we have?

—Tom Ehrich

What you do not want done to yourself, do not do to others.

—Confucius

What is hateful to you do not do to your neighbor. That is the whole Torah.

—Hillel

SIGNIFICANCE

An inconvenient truth is that the error can be within us. Our attitudes or dispositions may prevent us from accepting information gathered by appropriate procedures. We may deny or distort information that is contradictory to our beliefs. We may also falsely assume that the information provides absolute proof or disproof of a belief or action. The denial of contradictory information and the misinterpretation of the absoluteness of the conclusion from research contribute to significant errors that continue in everyday life.

Another concern is unethical procedures used to gather information. These unethical procedures can result in physical, emotional, or financial harm to the participants in research. Consumers of the information may also suffer harm by making erroneous decisions based upon information that was not ethically gathered or provided.

An old philosophical question is, Do the ends justify the means? In gathering information in research, this question has a clearer answer. If the end is to reduce errors in making decisions, then using unethical means to gather information may actually increase errors in decisions.

This chapter presents appropriate dispositions and ethical procedural safeguards for gathering information, with examples of the consequences of inappropriate dispositions and unethical conduct.

LEARNING OBJECTIVES

1. Accept that there may be errors in our beliefs.
2. Accept that our beliefs can change as factual evidence changes.
3. Accept that there may be errors in our behaviors.
4. Accept that our behaviors can change as factual evidence changes.
5. Accept that research is a process for continuous learning in identifying accurate and inaccurate evidence.
6. Be able to acknowledge an error in the evidence and discontinue erroneous beliefs and behaviors.

7. Accept that the most important ethical concern in research is to avoid any type of harm to the participants.
8. Understand that harm may be physical, financial, psychological, or in any other form in which the participants are adversely affected.
9. Understand that the researcher is ultimately responsible for any harm to the participants.
10. Understand that procedures can minimize any potential harm to participants and include these protections:

    a. Participants are provided with full disclosure of the methods used in the study.
    b. Participants must give consent to participate after being informed of the methods used in the study.
    c. Participants cannot be harassed or intimidated to participate.
    d. Discrimination is to be avoided in the selection of participants.
    e. Participants may refuse to participate without recrimination.
    f. Participants have the right to withdraw from participation at any time without recrimination.
    g. Participants have protection of confidentiality, as results of the study will not reveal the personal identity of any participant.
    h. Participants have the right to learn of the results of the study.

11. Understand that full disclosure of the procedures used in the study is required.
12. Understand that sufficient data must be reported and available to permit verification of the results by other researchers.
13. Understand that it must be accurately reported if a secondary source was used as a reference for a primary source.
14. Understand that citations are to be included for the scholarly work of others.
15. Understand that the researcher reports dual relationships when financially supported by a company or agency with a stake in the results of the study.
16. Understand that the researcher must report any potential conflict of interest that may bias the conduct of the study, including personal, professional, or financial interests.
17. Understand the potential harm when withholding a treatment or using a placebo treatment.
18. Understand that retraction is required when errors are identified after publication.

PREREQUISITES

1. Understand the definition of an error.
2. Accept that there may be errors in the evidence provided by research.
3. Be willing to accept the research procedures of the Scientific Method as one way of gaining knowledge.
4. Understand the sequence of steps in the Scientific Method.
5. Be willing to learn about research procedures in the Scientific Method.
6. Be willing to avoid erroneous application of the Scientific Method research procedures.
7. Accept that research provides only tentative answers with some uncertainty rather than absolute answers.
8. Demonstrate a disposition to avoid harm to participants when gathering and interpreting evidence.

CONCEPTS

Unlike most research principles textbooks, this textbook also addresses the attitudes, beliefs, and dispositions that are prerequisite for successfully applying research principles. Our experience is that if we approach research principles with a positive attitude or disposition, we shall increase our engagement and have a more successful learning experience. Try it: it will only hurt for a little while, and it is good for us. We could take two teaspoonsful with breakfast.

Each of us has given thought to our purposes in life and the world around us. Daily life provides us with new experiences and information. To explain the world and gain a sense of predictability and control that reduces the anxiety of being wrong, some of us retreat to the false certainty of simplistic thinking.

Some only assimilate new information to confirm existing beliefs and contradictory information is ignored or dismissed. These persons overvalue the implied certainty provided by traditions, rules, myths, faith, rituals, ideology, superstitions, dogma, common sense, magical thinking, or conspiracy ideas. Wow, that sounds like a menu of fast-food thinking! Can we go to a drive-through window and order a large dogma, a small myth, a supersized conspiracy theory, and an order of fries?

Some of us are less anxious about being uncertain or wrong and consider beliefs as tentative. These persons will accommodate or change their beliefs and behaviors when presented with new and even contradictory facts.

If we are looking for certainty, then research will only provide disappointment, because researchers are unable to prove or disprove anything on an absolute basis and must humbly make tentative conclusions based upon probability. Certainty is the sin of research and science, but the virtue of dogmatism. Scientists may accept that conclusions are tentative. Dogmatists do not acknowledge their errors and conclude that their conclusions are absolute. The strength of a belief does not indicate the accuracy of the belief.

The willingness to change beliefs as facts change is a fundamental disposition needed by scientists, because future research may expand, clarify, or contradict current understanding. Uncertainty or unpredictability is an important concept in our everyday life. Such is the life of researchers. What a group of flip-floppers!

An ancient Greek saying is: "Know thyself." What kind of a person are we? Do our beliefs determine the facts rather than the facts determining our beliefs? Are we unable or unwilling to change our beliefs when the facts change? Do we have the courage to admit when we are wrong and then change our beliefs as the facts change?

If we cannot admit to making errors and then commit to correcting the errors, then what hope do we have as a society? If that is your attitude or disposition toward the use of new information obtained from the research, then read no further and get a refund for the book. This textbook or any other textbook about research procedures will be of no use to you. You will have ignored the dynamic possibilities offered by research to increase our understanding. However, you will have the smug satisfaction that you already know everything.

Congratulations! Would you prefer your right or left profile on Mount Rushmore? Even the pigeons will be in awe of you and would not dare to land on your profile.

Certain research problems or questions that we may investigate involve research that could potentially embarrass or harm those people who are the participants in our research. What do we do? Do we say no pain, no gain, and continue with potentially harmful research? Do we change how we conduct research? If we change how we conduct research, will we be able to collect the data necessary to answer our question?

Guidelines or procedural safeguards are put in place to avoid harm to participants, and most institutions have a process to review these safeguards before permitting research. Here is a description of these safeguards:

Ethical Procedural Safeguards

1. Participants must be informed of the purpose and potential risk in the study.
2. Participants must give consent to participate.
3. Participants must not be coerced, intimidated, or harassed into participation.
4. Participants must not be discriminated against in selection.

5. Participants are free to refuse to participate without reprisal.
6. Participants have the right to withdraw from the study at any time without recrimination.
7. Confidentiality is protected so that the identity of the participants is not revealed.
8. Participants have an opportunity to learn of the results of the study.
9. Participants must be of legal age to consent to participate. A parent or guardian must give permission for a minor to participate.
10. Use of deception in conducting research requires approval by an appropriate person or committee in the agency. There must be a compelling reason for any deception.

We know we also have ethical concerns after conclusion of the research study. It may seem obvious, but researchers should not fabricate or make up the results, falsify or change the results, or plagiarize or take credit for the work of others. Fabrication, falsification, and plagiarism actually happen even in research published in peer-reviewed journals. The integrity of research depends upon the ethical standards of the researchers. For federally financed research projects, the Office of Research Integrity monitors ethical issues.

Negligence is an important ethical concern in the implementation of research procedures, data analysis, or reporting of results. Failure to follow accepted research principles, as discussed in this textbook, can contribute to erroneous conclusions that have profound adverse effects.

Many situations in research may present an ethical dilemma and a potential ethical failure. Is it ethical to give some patients a placebo (a false or meaningless treatment, like a sugar cube) or deny an effective treatment to see how their behavior compares to patients who receive an actual potentially effective treatment?

The researcher retracts published research after identification of errors or negligence. This may happen after the researcher has further reviewed the conduct of the study and provides an ethical way to alert others regarding the accuracy of the research. The researcher must also keep the original data of the study for a period to permit other researchers to verify the results.

Many situations require us to make decisions with the honesty of our own thoughts and actions and with consideration for the potential harmful effect upon others. We have all faced situations in which we have lied. Researchers face these same temptations and must abide by the ethical standards in conducting research. What will we do?

Did we harm another person because of our lie? Are we able to go through an entire day telling the truth? Do we exaggerate or minimize an action without considering it a lie? Here are some phrases that we hear in everyday life to justify untruthful actions: "I hope that I don't get caught";

"Of course I will respect you in the morning"; "Did you read the fine print?"; "It's only a fib"; "No one will ever know"; "Buyer beware!"; "How can we put lipstick on this pig?"; "Everyone else cheats."

The courage to make the right choices to avoid harm to others is part of the potential nobility of humanity, the core principle of an ethical life. Unfortunately, we sometimes deceive others and ourselves or do unto others what we would not want done unto us. The cowardice in making incorrect choices that can harm others is part of the potential reality of humanity. We all face these conflicting choices, but the hope is that we make choices that avoid harm to others.

The goals of conducting research are to identify the most accurate and truthful information while simultaneously avoiding harm to those involved in the research. Before we actually conduct research, we must be committed to truthful investigations using the most appropriate methods, accepting the results of those investigations even if contrary to our existing beliefs, and avoiding conducting investigations that can harm those involved in the research.

COMMON ERRORS

1. Not considering information that is contradictory to personal beliefs
2. Attempting to maintain a sense of control and reduce the cognitive dissonance between personal beliefs and the factual evidence by using any of the following statements or actions meant to deny, evade, minimize, obfuscate, or distort contradictory information:

    a. "That is your opinion."
    b. "I do not believe you."
    c. "Not all scientists agree."
    d. "You are a(n) [fill in a label with an ____ist, ____ism, or ____ic]."
    e. "You can't prove that."
    f. "Research proves . . ."
    g. "It is common sense."
    h. "Everyone does."
    i. "You can make research show anything that you want."
    j. "You can make statistics show anything that you want."
    k. "We always do it this way, it's tradition."
    l. "You're just trying to trick me."
    m. "You're just making that up."
    n. "I don't have time to listen to you."
    o. "That's just a theory."
    p. "I know what you really think."

shore. If the earth were round, then we would only see the part of the hull not blocked by the curvature of the earth as it moved farther from shore.

It may seem silly to us that people thought the earth was flat. For how many centuries did this belief prevent exploration because of fear of falling off the edge of the earth? Many people died for believing that the earth is round rather than flat even though the evidence is consistent with the explanation of a round earth. According to the Pew Research Center, 18% of us continue to believe that the earth is flat and that there really are "four corners of the earth."

In today's competitive job market, employers are seeking those who display the same attitudes of researchers: continuous learning and a willingness to change beliefs as the facts change. Failure to incorporate these attitudes and change our beliefs and actions as the facts change is among the most serious errors that we make. Every human has to change in many ways from birth to adult life. Have we stopped growing and changing? If so, then why?

- *USA Today* reported that Richard Carmona, US surgeon general from 2002 to 2006, stated that factual evidence indicated that abstinence-only sex education programs were ineffective. He further stated that the Bush administration forced him to emphasize abstinence-only sex education to appease religious conservatives.
- Senator James Inhofe of Oklahoma has repeatedly denied the existence of climate change caused by humans and has variously stated that not all scientists agree that climate change exists. The senator then says that he has a list of hundreds of scientists who deny climate change. How many Nobel scientists were on the senator's list?
- *USA Today* reported that evangelical radio minister John MacArthur wrote that the environmental movement is wrong to try to preserve the planet because "the Lord is going to destroy it." (We guess that also means we don't have to wash the dishes after dinner, since we will not have to preserve them. We always hated washing dishes, and now we have a reason not to.)
- CNBC.com quoted Rick Santelli's August 2013 comments about the rate of inflation reported by the government as follows: "Listen. I don't believe the government's calculations. There, I said it. I don't have better numbers, I have common sense." Is it an error to ignore information and rely upon personal beliefs?
- *The New York Times* reported that even after Senator Rand Paul was told that government employment declined under President Obama, he continued to insist that "growth of government is enormous under President Obama." He refused to change his belief when confronted with factual data showing that federal government employment had declined under the Obama administration compared to the Bush administration.

- The medical journal *The Lancet* retracted published studies by Dr. Andrew Wakefield, indicating a link between the vaccine for measles, mumps, and rubella (MMR) and autism. Wakefield acted in an unethical manner by falsifying the data. Many nonscientific TV talk show hosts use this falsified data to support stopping vaccination.
- Public Policy Polling discovered that 54% of Americans either agree or are not sure that there is a link between childhood vaccines and autism. Is there any harm if children do not receive vaccines against these diseases? Many parents have refused to have their children receive the MMR vaccine.
- *Family Planning Perspectives* reported that the notion that sex education in school leads to promiscuity among teenagers is erroneous. The research found no significant differences in age of first sexual experience or number of reported partners. Are there any consequences of maintaining an erroneous belief about sex education and promiscuity?
- Brian Martinson reported, in the British journal *Nature*, the results of an anonymous survey of 3,200 US scientists in which 33% admitted to breaking rules in conducting research within the past three years. The ethical errors range from fraud to improper relationships with the research subjects. Fifteen percent of scientists stated that they had changed the methods or results because of pressure from the financial sponsor of the research. Falsified information is an ethical error that causes harm to the consumers of the information.
- The United States Department of Commerce reported on February 25, 2011, that deeper spending cuts by governmental agencies slowed economic growth during the last quarter of 2010 and could reduce growth in the economy to 2.8% from 3.2%. The solution offered by the Republican Party during the budget discussions of 2011 to increase economic growth and reduce the deficit was to cut more spending. This is an error of ignoring disconfirming data.
- CNN.com reported that RMC Research Corporation, the government contractor, developed and evaluated the $1 billion Reading First program. The inspector general of the Government Accountability Office found that federal officials intervened to influence decisions by the states regarding which reading programs to adopt. RMC also hired subcontractors with financial ties to the reading products used in Reading First. Is there an ethical error of a potential conflict of interest?
- Here is a classic example of ethical errors: The Tuskegee syphilis experiment involved observing men to determine the natural progression of syphilis, with no informed consent given for participation. None of the men received an effective treatment. The men were never told that they had syphilis, and they were never given any treatment other than a placebo. Is there any harm here?

REDUCING ERRORS

Errors may be inevitable. Prevention or avoidance of errors is preferred to the need to correct errors after they have occurred. Numerous appropriate procedures described in this textbook may reduce, but not eliminate, the possibility of errors.

As individuals and as a society, we must be open to the possibility of errors and constantly monitor for errors. In accepting this possibility, we must also humbly accept that the results of any research provide only tentative conclusions. This is an important disposition to adopt, as it prevents the erroneous assumption that research provides absolute proof or disproof of our beliefs and decisions. The interpretation of research information is in terms of probability so that we acknowledge the need to continue to gather information to obtain a more complete understanding. Think of how this disposition can prevent many conflicts.

Unethical behavior in gathering information in research contributes to both erroneous decisions and harmful effects for the participants. Following the ethical procedural safeguards described in this chapter can help minimize erroneous and harmful effects. Avoiding harm to the participants in research studies and to the consumers of information is an important consideration in all research.

Prevention of all errors may not be possible. Reducing the possibility of errors is feasible. The disposition to openly examine contradictory information and change beliefs and behaviors as accurate information becomes available is paramount in reducing errors. This disposition also reflects an ethical consideration to seek and use accurate information as the basis of personal beliefs and actions.

BEST PRACTICES

- Follow ethical safeguard procedures presented in this chapter.
- Follow ethical procedures involving full disclosure in conduct of study, dual relations, and conflicts of interest.
- Change beliefs and behaviors as more factual information is obtained.

INTERRELATION TO OTHER TOPICS

This chapter indicated that, to reduce errors, an appropriate disposition is to modify beliefs and behaviors when presented with accurate information. A commitment to ethical procedures was also included.

Ethically, we must also be aware of potential harmful effects on participants in a research study or on the consumers of the information. Ethical safeguards include procedures to follow in gathering and interpreting information.

The next chapter provides an understanding of the nature of the problem for which we gather evidence. The problem involves resolving the difference between our current state of knowledge or behaviors and a desired state.

The next chapter includes several important concepts and procedures regarding developing the problem to investigate. The first is reviewing relevant literature. The second is understanding the importance of a theory and how information supports or refutes the theory. The third is how to construct or frame a problem to guide the subsequent research activities. The fourth is the development of a testable research hypothesis and the concept of the null hypothesis. Inaccurate or inappropriate procedures in any of these areas can contribute to erroneous decision making.

Chapter Three

Problem Development

"Stepping Stones on a Yellow Brick Road"

The greatest challenge to any thinking is stating the problem in a way that will allow a solution.

—Bertrand Russell

It is a test of true theories not only to account for but to predict phenomena.

—William Whewell

No idea is so antiquated that it was not once modern. No idea so modern that it will not someday be antiquated.

—Ellen Glasgow

The investigations of others can be a substitute for personal experience.

—Anonymous (this quote is not by us this time)

Knowledge is of two kinds. We know a subject ourselves, or we know where we can find information upon it.

—Samuel Johnson

All questions are answerable, but we must first identify the exact question that we want to answer, Plato.

—Socrates (trust us: we are so old that we were there taking notes on our stone tablets as Socrates said this to his student, Plato)

The difficulties and disagreements, of which history is full, are mainly due to a very simple cause: namely to the attempt to answer questions, without first discovering precisely what question it is which you desire to answer.

—George Moore

Every man is fully satisfied that there is such a thing as truth, or he would not ask any question.

—Charles Peirce

The shrewd guess, the fertile hypothesis, the courageous leap to a tentative conclusion—these are the most valuable coin of the thinker at work.

—Jerome Bruner

The great tragedy of Science: the slaying of a beautiful hypothesis by an ugly fact.

—Thomas Huxley

Here is the conclusion on which I will base my facts.

—Adlai Stevenson

A hypothesis is whether something is true or false. Evolution is more than a hypothesis because there is proof.

—Monsignor Gianfranco Basti

SIGNIFICANCE

The problem that we investigate determines the type of information we need and the procedures we use to gather that information. The problem must be stated in a manner that is both capable of investigation and that permits the development of a question that can be answered and a hypothesis that can be supported or refuted. Inaccurate or inappropriate development of the problem can result in erroneous selection of all subsequent research procedures and in an erroneous decision.

Multiple types of errors made in the development of the problem are described in this chapter. They include inaccurate or erroneous sources of information, vague or erroneous operational definitions of terms and concepts, erroneous understanding of a theory, erroneous alignment of the type of conclusion with the type of question investigated, and failure to develop a testable hypothesis before conducting research.

These errors are present in our daily lives. Identifying these errors can assist us in making informed decisions. This chapter presents appropriate procedures in problem development and examples of errors made in developing a problem to investigate.

LEARNING OBJECTIVES

1. Understand that a research problem to investigate is identified from personal experience or from professional literature.
2. Understand that a problem states the difference between the current and a desired status.
3. Understand that a problem must also state a desired solution.
4. Understand that a problem must be capable of investigation.
5. Understand that a problem must be significant enough to investigate.
6. Understand that a problem investigated addresses one of these purposes:

    a. Extend limits of previous evidence.
    b. More precisely clarify current evidence.
    c. Correct errors in current evidence.
    d. Examine other potential explanations for the evidence.
    e. Determine whether intermediary variations affect the evidence.
    f. Replicate previous research to verify current evidence.
    g. Test a theoretical prediction of evidence.

7. Understand that a theory provides an organized set of explanatory assumptions and is capable of support or refutation by factual evidence.
8. Understand that a problem states a theoretical foundation.
9. Understand how the literature review supports the purpose for investigating the problem.
10. Identify differences between primary and secondary sources of information in the literature.
11. Understand the need to use peer-reviewed sources of information.
12. Understand the need to conduct a comprehensive and current review of the literature.
13. Understand the importance of using research evidence and not opinions from literature.
14. Understand the importance of making an analysis of previous literature to provide a theoretical support for the problem investigated.
15. Understand computer-based strategies to locate previous relevant literature.
16. Understand appropriate ways to cite references from the literature.
17. Understand how to construct and use operational definitions.
18. Understand the different types of research questions and research designs:

a. Descriptive
b. Comparative
c. Correlational
d. Predictive
e. Causal-comparative
f. Ex post facto
g. Quasi-experimental
h. Experimental
i. Single-subject
j. Action
k. Ethnographic
l. Mixed-methods

19. Understand that a hypothesis is a statement of an expected outcome of the research.
20. Understand that a hypothesis must be capable of testing for support or refutation.
21. Understand that a hypothesis is made a priori and before the study.
22. Understand that the hypothesis should operationally define:

 a. Subjects or persons in the study
 b. Evidence to be collected
 c. Treatments to the persons in the study
 d. Anticipated results
 e. Probability that result is due to chance

23. Understand that the research hypothesis states either (a) predicted direction of the difference between groups (one-tail) or (b) difference regardless of the direction of the difference (two-tail).
24. Understand that the null hypothesis is an alternative statement that indicates that what was done in the study produced no difference or the difference was in the opposite direction from the evidence predicted by the research hypothesis.
25. Understand that a statistical level of probability determines acceptance or rejection of the null hypothesis.
26. Understand that there are four outcomes from testing a research hypothesis:

 a. Correctly rejecting the null hypothesis
 b. Correctly keeping the null hypothesis
 c. Incorrectly rejecting the null hypothesis (Type 1 error)
 d. Incorrectly keeping the null hypothesis (Type 2 error)

27. Understand that keeping or rejecting the null hypothesis is based only on statistical probability and not the practical significance or consequence of a Type 1 or Type 2 error.

PREREQUISITES

1. Understand the definition of an error.
2. Accept the importance of reducing errors.
3. Demonstrate a disposition to continuous learning in identifying erroneous information.
4. Demonstrate a disposition to change beliefs when facts change.
5. Demonstrate a disposition to apply ethical standards in the conduct of research and the interpretation of results.
6. Demonstrate a disposition to accurately report the conduct of a research investigation.
7. Demonstrate a disposition to avoid harm to participants when gathering and interpreting evidence.

CONCEPTS

Perhaps each of us has so many problems that we often do not know where to begin or what to do to solve the problem. We will spend time agonizing over how the problem came to be and what solutions might be available. We will talk to others, read books, or search the Internet to get some ideas for solutions to our problems. We will accept some of those ideas and reject other ideas. Do we believe that only true information can be on the Internet? It is important that we do not make an error and choose the incorrect solution.

We may identify a problem to investigate based upon our personal experiences or introspection. Just listen to common phrases that we might use: "I just came up with a crazy idea"; "I wonder why that happened?"; "I wonder what would happen if . . . ?"; "What is the best choice?"; "How do I make this better?"; "What do most people do?"; "I heard it through the grapevine."

For researchers, the research problem provides a direction for the application of all subsequent activities and research procedures. The problem must be significant enough on a personal, group, or societal level to justify the research effort. It takes time to plan and implement accurate research, and there may be financial costs involved.

If we already know everything, then we may not see a need for research (we keep forgetting if our left or right profile will be on Mount Rushmore). Research gathers facts to reduce the uncertainty of knowledge. Sometimes, the only way we know if a potential solution works is to try it and observe the results. The specific research problem investigated should address one of the following purposes:

Purposes for Conducting Research

- Extend the limits of previous research to new or additional situations, people, and measures.
- Clarify current knowledge with more precise information.
- Correct errors in previous research.
- Delete or control other potential causes.
- Determine if intermediary amounts or values cause same or different results.
- Replicate previous research to verify and refine earlier results.
- Test a theoretical prediction.

Developing a clear statement of a problem to investigate may be challenging. Can we identify a problem that we see in society or in our personal lives? Is it significant? Can we identify one of the above purposes to help us to identify the information that we need? The professional literature may help direct us to the type of research problem investigated.

Everyone lives with a theory or theories about the causes or reasons for many of the actions that we see around us. A scientific theory provides an organized set of assumptions that explain and predict and is capable of support or refutation by verifiable factual data. What are our theories about methods of teaching, ways to raise children, size of the government, interpersonal relations, business regulation, or climate change? Are our theories capable of support or refutation by accurate verifiable data? If not, then why not? Do we ignore contradictory information (did we talk about this in the last chapter)?

The common misunderstanding is that a theory is just a set of personal beliefs and assumptions that a person holds. Often, this misunderstanding is dismissive of explanations other than a person's personal views. We see much discussion and dismissal of Darwin's theory of evolution as "just a theory." However, decades of verifiable factual data—and the inclusion of DNA and DNA mutations as active agents in the evolution of a species—support this theory.

In contrast, intelligent design is a belief system that is not capable of support or refutation by verifiable facts, and provides a false sense of certainty that humans have always been as we now are and that humans will not change. Denying evolution eliminates the uncertainty of change. The average weight of a college football lineman in the 1930s was approximately 190 pounds; in 2013, it is approximately 300 pounds. (Is it a 15-yard penalty for using a fact to discredit intelligent design?)

We live in a world with so much available information from a great variety of sources, such as TV, radio, Internet, newspapers, professional literature, and interaction with others. Finding accurate information from any of these sources can save us much time and personal effort, as we can learn from the earlier efforts of others.

Many professional computer-based search engines help us to obtain relevant information that meets the various purposes for research. Search engines affiliated with a professional academic society or a governmental agency usually have more accurate information than general search engines. The strategy for using these search engines is to identify keywords related to our research question. Keywords often are in the first page of published literature.

The speed of these computer searches provides the opportunity to try different keywords or combinations of keywords. Reference librarians are of great assistance in beginning our review of the literature.

Conducting an appropriate review of the professional literature is an important research procedure that supports and directs additional procedures. Peer-reviewed journals are the most accurate source of information. Studies published in these journals were determined to have acceptable application of research procedures to support the results. But publication in a peer-reviewed journal does not guarantee accuracy, as there may still be errors in the application of the research procedures.

The original or primary source of information, rather than a secondary source of someone else's interpretation of the original material, should be used. Something does get lost in translation or interpretation. How many times have we spread rumors or what we thought someone said, rather than check with the original source? (Captain Kirk never said, "Beam me up, Scotty," and Humphrey Bogart never said, "Play it again, Sam.")

An appropriate review of the literature should include current, comprehensive, peer-reviewed, and primary sources of information of the topic investigated. The review should permit an analysis of the strengths, weaknesses, and limitations of the findings of previous research rather than just a summary or the opinions of previous researchers. The review can help us to identify errors in previous research, suggest new measures or methods to try, and suggest theoretical foundations.

A review of the relevant literature may also identify potential explanations for conflicting results in previous research. Important reasons for these conflicting results may include erroneous application of different research procedures, and different studies may use different operations to measure an identical term or characteristic.

The specific operations provide useful definitions of the term or characteristic; this is an important part of research, because the definition provides a clear and consistent communication regarding how a term or characteristic is measured. It is important to examine the operations or how different researchers measure the terms and characteristics investigated. For example, what is "intelligence"? Is it something that we can see, or is it just the score on an intelligence test? We have many different intelligence tests, so which one(s) are used, and are those the most appropriate?

Both individuals and researchers can get into arguments based upon having a different definition of a term or characteristic. A great way to start an argument is to ask someone for their operational definitions of these terms: "love," "socialism," "Christian," "free market," "conservative," "entitlement," "bureaucracy," "government control," "at-risk."

An accurate citation of all reference material used in the development of the research investigation should be included. The *Publication Manual of the American Psychological Association* is a commonly used guide to the accurate citation of references.

Each of us has questions about many different topics and at different times in our lives. Try talking to a 3-year-old and you will hear an endless series of the same question: "Why?" Asking questions helps each of us to learn and to understand how the world around us functions. That wonderful curiosity in asking questions shown by young children diminishes as we become older. We may believe that we already know everything, or, perhaps, we are afraid to admit that we do not know everything or anything.

A necessary task that all of us face is to clarify thinking about the question and the answer that is sought. The type of question that we ask is the beacon that directs us in the activities that we undertake and in obtaining the necessary information. The quality of the answers we obtain depends upon the appropriateness of the research principles applied.

The following presents a brief summary of the types of questions asked in our personal lives and by researchers. Try to frame questions as one of these types.

Types of Research Questions

- Descriptive questions try to identify the current of a group or an event. For example: How many people did this? What is the average of the group? What is the range of scores in the group?
- Comparative questions try to determine whether there are differences between groups on one or more dependent variables. Any obtained difference does not imply a causal relationship. For example: Do students of experienced teachers have different rates of attendance compared to student of beginning teachers? Do male students college students have different graduation rates compared to female students?
- Correlational questions examine relationships between different characteristics. For example: Is there a relationship between sales and advertising

increases? Do my test scores increase as I study for more hours? Does my blood pressure decrease as I spend more time exercising?

- Predictive questions investigate whether scores on a characteristic predict scores on other characteristics. For example: Are college grade point averages predicted by SAT test scores and time spent studying? Do spending cuts predict the unemployment rate? Do hours spent in premarital counseling predict eventual years of marriage?
- Causal-comparative questions explore whether a naturally occurring preexisting current characteristic affects another characteristic. For example: Do girls do better than boys in reading tests? Do older persons have fewer car accidents than teenagers do? Are rich people happier than poor people?
- Ex post facto questions try to determine the causes of a past event. For example: Did students who had a first-year mentoring program have a higher graduation rate than students who did not have that program? Did prisoners who received job training have a lower recidivism rate after release from prison? Did children who received vaccines have lower rates of disease than children who did not receive the vaccines?
- Quasi-experimental questions examine whether groups that will receive a different manipulation of a characteristic will perform differently on another characteristic. However, there is no assurance that the groups were equal at the time of the manipulation. Questions might include the following: Will sales be higher when a group receives an advertisement with a celebrity endorsement than when a group receives an advertisement without a celebrity endorsement? Will students have higher test scores in an online class than students in a traditional class?
- Experimental questions try to determine whether different direct manipulations of a characteristic will cause a different result or effect in groups. There is some assurance that the groups are equal to begin with because of random assignment to the groups and the control of other explanations. Questions might include the following: Will patients randomly assigned to receive the new medication have a faster recovery than those patients randomly assigned no medication? Will voters randomly assigned to receive a political attack ad cast fewer votes for a candidate than voters not randomly assigned to receive the attack ad? Will students randomly assigned to receive tutoring have higher achievement test scores than students not randomly assigned to receive tutoring?
- Single-subject questions study an individual rather than groups and compare how the person did before, during, and after the direct manipulation of a characteristic. For example: Will rewards reduce inappropriate behaviors? Will there be an increase in the productivity of an employee after a promotion?

- Action research questions study the immediate effects of treatments as a guide to improving an intervention. For example: Will a time-out for students create an immediate improvement in class behavior? Will posting a sale price in a store for the next few minutes create increase sales?
- Ethnographic research studies the cultural patterns of a group in the naturally occurring context by use of observation, interviews, and examining artifacts. Questions might include the following: What are traditional patterns of family organization in this culture? What are the cultural practices regarding marriage in this culture?
- Mixed-methods research is a hybrid that combines descriptive and experimental or quasi-experimental research processes. The purpose of this hybrid approach is to provide a comprehensive understanding beyond what one method of study provides. Questions might include the following: What are effective ways to present ads for car sales? Do the ways that football coaches interact with players affect the performance of the players?

We go through life with many hypotheses as to what we think will happen or how things work. Sometimes we are right, and sometimes we are wrong. The important point is that we gather information that allows us to determine the support for our expectation.

Researchers go through that same process of developing hypotheses or expectations of what will happen in certain conditions or situations. These expectations are in advance (a priori) of an investigation rather than after the investigation (ex post facto). Stating hypotheses in advance makes researchers more careful, because the risk of refutation contrasts with taking no risks with ex post facto hypotheses. A hypothesis should have at least these components:

Components of a Hypothesis

- subjects or persons in the study
- measurements or evidence gathered to test the hypothesis
- what is done to the subjects in the study
- anticipated results or prediction
- probability that the results or predictions are due to chance

Accurate information collected and analyzed by appropriate methods and procedures tests the hypotheses. A one-tail hypothesis predicts the direction of the expected difference between observations, such as "There is global warming where the current temperatures are higher than past temperature." A two-tail hypothesis only predicts a difference, without regard to direction, such as "There is climate change." Whatever the hypothesis that we develop, it must be capable of testing with verifiable data that can support or refute it.

The research hypothesis is never directly tested. Support for the research hypothesis comes only from inference. A separate null hypothesis states that the results of a study are due to chance. When the results are unlikely due to chance, the null hypothesis assumed incorrectly and the research hypothesis inferred correctly. This can be a correct conclusion, but if the null hypothesis is actually correct, then this error is a Type 1 error.

If the obtained results are likely due to chance, then the null hypothesis assumed correctly and the research hypothesis inferred incorrectly. This can be a correct conclusion, but if the null hypothesis is actually incorrect, then this error is a Type 2 error. We have to determine whether a Type 1 or Type 2 error is the most serious, because the greater the chance of one type, the smaller the chance of the other type.

Forming an appropriate hypothesis for research investigation is an important task and is dependent upon an understanding of the limitations and errors of previous research, as well as an accurate understanding of all of the research procedures discussed in this textbook. Remember, the hypothesis must be capable of support or refutation after the collection of facts. The hypothesis states appropriate operational definitions of the persons, measures, treatments, methods of statistical analysis, and level of acceptance of the probability of chance.

COMMON ERRORS

1. Not indicating a difference between the current and desired state
2. Selecting a problem that is not capable of investigation through obtaining verifiable facts
3. Not indicating any of the following purposes for conducting the research:

    a. Extending limits of previous research with additional variables or range of values of variables
    b. Obtaining more precise information
    c. Correcting errors in previous research
    d. Deleting or controlling other potential causes
    e. Determining if intermediary amounts or values cause the same or different results
    f. Replicating research
    g. Testing theories

4. Not stating a theoretical foundation for the problem investigated

5. Not using primary sources of information
6. Using information that is not from peer-reviewed sources
7. Inaccurate citation of references
8. Conducting an incomplete review of the literature
9. Conducting a review of the literature that is not current
10. Using references from sources known to have bias and inaccurate methodology or conclusions
11. Not identifying the keywords used in a computer-based search of the literature
12. Providing a summary rather than an analysis of the literature reviewed
13. Inaccurately analyzing the conclusions of the studies reviewed
14. Not providing operational definitions for characteristics examined
15. Not indicating in a descriptive question the characteristic on which statistical methods summarize the group performance
16. Not indicating in a correlation question the characteristics that are examined for relationship
17. Not indicating in the correlation question the expected strength and/or direction of correlation
18. Inaccurately indicating that a correlation question can answer a cause-effect relationship
19. Not indicating in a prediction question the characteristics of the predictors and/or the criteria
20. Not indicating in the hypothesis an assumed degree of accuracy in a prediction question
21. Inaccurately indicating that a prediction question can answer a cause-effect relationship
22. Inaccurately indicating that the naturally occurring preexisting difference between groups on a characteristic in a causal-comparative question represents direct control in establishing cause-effect relationships
23. Inaccurately indicating in the hypothesis that the past results obtained from an ex post facto question represent direct control in establishing cause-effect relationships
24. Inaccurately indicating in the hypothesis an initial equivalence between groups in a quasi-experimental study where there is a lack of random assignment
25. Not examining the change between before, during, and after a manipulation or intervention in a single-subject question
26. Not specifying hypothesized outcomes a priori
27. Inability to obtain evidence to test for confirmation or refutation because of lack of accurate operational definitions for one or more of the component parts of a hypothesis
28. Failing to include one or more of the component parts of a hypothesis
29. Inability to test the hypothesis due to ethical concerns

EXAMPLES

- *The New York Times* reported a widespread industry practice of ghostwriting medical research studies published in academic journals. Merck used ghostwriters to draft many research studies for its drug Vioxx and then paid doctors who did not conduct the research to put their names on the reports. Merck had to pay $4.85 billion in a later settlement because the drug increased heart attack risk and deaths. Research published in peer-reviewed journals is no guarantee of accuracy. The errors may be in original data or with the original authors of the report. (It is a bitter pill to swallow when we have doubts about the accuracy of medical research. Oh, well—perhaps they will get a taste of their own medicine.)
- *The New York Times* reported that J. Hendrik Schoen, Bell Labs scientist, published 17 research studies in physics based upon fraudulent data even though peers in scientific journals could not identify any concerns. The important assumption is that peer review can identify errors in research procedures or analysis, but the assumption is that the data are honest. The errors and the dishonesty in the data may be difficult to identify even with professional reviewers. (Our guess is that Schoen will not use his frequent flier miles to go to Stockholm for the Nobel ceremony.)
- *The Plain Dealer* reported that Richard McKenzie is a college professor and a self-made millionaire (we never thought that we would see "professor" and "millionaire" in the same sentence). McKenzie included these simple and universal "keys" to becoming wealthy: hard work, honesty, study, and a little frugality. Are there any potential errors in how different researchers might operationally define these terms to test this hypothesis? What is a "little frugality"?
- CNN.com reported that Tim White, a paleontologist at the University of California, Berkeley, claimed that a 4.4 million-year-old skeleton supports the theory that humans and chimpanzees evolved separately from a common ancestor. White asserted that further research is required to determine the precise time at which chimps and humans split from their common ancestor. What precise information is required, and how precise? (Wonder if some old skeletons in a closet can answer this question. What skeletons are in our closets?)
- *The Plain Dealer* reported that Jonathan Gruber of the Massachusetts Institute of Technology identified dozens of studies that found a link or correlation between religion and health, higher incomes, and other behaviors. Gruber stated that he did not know why there is a link. Can this problem control for some potential causes? Is it an error to conclude that religion is a cause of better health and higher income? Is it an error to conclude that religion is a result of better health and higher income? (Probably need to pray for guidance in answering the link with religion.)

- *The New York Times* reported that economists Carson Reinhart and Kenneth Rogoff had unintentional errors in their spreadsheet data regarding sovereign debt and economic growth. The erroneous data were widely used by many to justify economic austerity. Examination of the original spreadsheets rather than secondary sources identified the errors. What have been the consequences of developing economic policy based upon interpretation of erroneous economic data?
- Blogs routinely report conspiracies, misinformation, and fabrications, as there is no peer review to check for accuracy. As reported in *The New York Times*, Rush Limbaugh read excerpts from what he thought was Barack Obama's senior thesis in which there were critical comments about the Constitution of the United States. However, Limbaugh's source was secondary, a blogger who posted inaccurate comments attributed to President Obama, rather than from President Obama's actual original document. Limbaugh later admitted that the blog had erroneous information.
- Satoshi Kanazawa of the London School of Economics and Political Science theorizes, in *Social Psychology Quarterly*, an evolutionary basis of supernatural and religious beliefs. Those who were slightly paranoid and looked for danger or enemies where none existed had a survival advantage over less religious persons who did not suspect a danger or an enemy where one existed. (Good Lord, this is a touchy topic! Is religion the result of evolution? Have religious practices changed or evolved over the years? What information will answer the question?)
- *The Plain Dealer* reported a study by The Center for Effective Discipline that found that schools that paddle students more often are more likely to have lower test scores. Are low test scores a significant problem? Does paddling cause low test results? Are there erroneous cause-effect interpretations because of other potential explanations? Are there answers to these questions?
- *The New York Times* reported that Pat Robertson stated that God used Hurricane Katrina to punish the people of New Orleans for their sinful life. Is this an erroneous conclusion? The residents of low socioeconomic status had less access to transportation to leave. Does God like rich people better than He likes poor people? How do we test this hypothesis? What explanation is there for a lighting strike that destroyed a 70-foot statue of Jesus in Ohio? Is God the cause of only some things or of all things? (Is a lightning strike an act of God?)
- *The Plain Dealer* reported that Julie Zito of the University of Maryland said that approximately 150,000 children between the ages of 2 and 4 were taking stimulants, antidepressants, or antipsychotic drugs, even though none of the most common of these medications had approval for use in children under 6 years old. Research is required to test the hypothesis that the medications have positive effects on children of this age. Since there is

no approval of the medications for use in children, an ethical error prevents testing of the hypothesis.

- Try to test this hypothesis: "We rejoice in our sufferings, because we know that suffering produces perseverance: perseverance, character; and character, hope." This assumed causal link is from Romans 5:3–4, in the Holy Bible. Can we test this hypothesis? How do we operationally measure the concepts? Does it matter what amount or type of suffering? Can research examine the Bible to obtain evidence to support or refute the statements?
- A common phrase that represents an ex post facto hypothesis in everyday life is: "I just knew that it would happen." Have we ever used excuses after we did or did not do something? Was that an ex post facto hypothesis, or just a fabrication to give an excuse? (We would give a list of excuses that we have heard over the years, but we did not get back in time last night to make the list. We were fighting our way out of capture by space aliens.)

REDUCING ERRORS

Reducing errors requires the application of multiple, appropriate procedures in developing the problem to investigate. An important procedure involves the review of previous research and avoidance of inaccurate or fabricated sources of information. Use of peer-reviewed journals or information from reputable sources is essential to understand previous discoveries and to clarify the purpose and need for any further research.

Prior to gathering the information, the type of question and hypothesis are developed. Stating these in advance minimizes the potential biases or errors that may result from making interpretations after the fact. The hypothesis must be capable of support or refutation with the obtained information. Terms, concepts, persons, measures, and expected results need clear operational definitions to avoid errors based upon vague or confusing meanings.

Research must be capable of gathering verifiable information in an ethical manner. Ethical errors can result in fabricated or falsified information, and can create inaccurate results that prevent application to other situations. Decisions must be made with verifiable information to avoid the potential errors that arise from decisions based upon estimated or hypothetical results.

BEST PRACTICES

- Form a clear statement for the purpose of research study (e.g., extension, greater precision, correction of errors, alternative explanations, intermediary variations, replication, test a theory).

- Accurately identify the type of research question (descriptive, comparative, correlational, predictive, causal-comparative, ex post facto, experimental, single-subject, action, ethnographic, mixed-methods).
- Conduct a comprehensive review of literature that supports the purpose of the study.
- Provide precise operational definition of all terms.
- Form a testable hypothesis (e.g., subjects, treatments, measures or information collected, anticipated results, level of probability of chance in decision).
- State the hypothesis a priori.

INTERRELATION TO OTHER TOPICS

We have just identified how the purposes for investigating a problem are developed and supported from previous literature and theories. The problem and the research hypothesis should be capable of investigation to obtain verifiable evidence that supports or refutes the hypothesis.

The next chapter describes variables that are measures of characteristics used to obtain information in a study. Variables are categorized as independent, extraneous, or dependent. Independent variables determine a potential cause or predict an outcome. Extraneous variables are characteristics that may alter the determination of cause or the prediction and contribute to an erroneous conclusion. The dependent variable is sometimes called the outcome variable and is commonly assumed to be the evidence in a research investigation that supports or refutes the hypothesis.

Chapter Four

Variables

"The DNA of Research"

The present contains nothing more than the past, and what is found in the effect was already in the cause.

—Henri Bergson

Find out the cause of this effect.

—Shakespeare

The right of the people to be secure . . . and no warrants shall issue, but upon probable cause.

—Fourth Amendment, Constitution of the United States

As we grow older, the world becomes stranger, the pattern more complicated.

—T. S. Eliot

Life being all inclusion and confusion.

—Henry James

God hath chosen the foolish things of the world to confound the wise.

—The Holy Bible

He chose to include the things that in each other are included, the whole, the complicate, and the amassing harmony.

—Wallace Stevens

When one tugs at a single thing in nature, he finds it attached to the rest of the world.

—John Muir

Every cause produces more than one effect.

—Herbert Spencer

The cause is hidden, but the result is well known.

—Ovid

SIGNIFICANCE

Variables are measures of characteristics that can provide a range of categories or scores. The specific variables used and how they are measured can affect accurately supporting or refuting the hypothesis. These variables need operational definitions to permit careful measurement in gathering information.

Multiple types of errors occur in the selection of variables, and these errors have significant adverse consequences for the accuracy of decisions. Researchers may neglect to include variables that cause or that can predict an outcome. The variables that are used may not have a sufficient range to determine cause or predict with accuracy. An insufficient number of outcome variables may also limit the ability to detect any variation or inconsistencies.

Selection of variables is an important task in conducting research. The variables must be relevant as a potential cause or prediction and for the type of outcome. Some potential variables can modify any obtained cause or prediction and contribute to errors in decision making.

LEARNING OBJECTIVES

1. Understand the concept of a variable.
2. Understand the meaning of the concept of independent variable.
3. Understand direct or indirect manipulation of an independent variable.
4. Understand direct or indirect manipulation of an independent variable by type, amount, or presence or absence, single or multiple.
5. Understand the concept of multiple independent variables (factorial) simultaneously investigated in a study.
6. Understand the advantages of a factorial study compared to multiple studies with only one independent variable.
7. Understand that an independent variable that is used to predict is a predictor variable.

8. Understand that there may be multiple predictor variables simultaneously used in a study.
9. Understand the advantages of using multiple predictor variables compared to multiple studies with only one predictor variable.
10. Understand an extraneous variable.
11. Understand a mediating variable.
12. Understand a moderating variable.
13. Understand a confounding variable.
14. Understand a multi-collinear variable.
15. Understand a dependent variable.
16. Understand multiple dependent variables (multivariate analysis).
17. Understand the advantage of multivariate analysis compared to multiple studies with only one dependent variable in an investigation.
18. Understand a criterion variable in a prediction investigation.
19. Understand the advantage of multiple criterion variables in a prediction investigation.
20. Develop operational definitions of dependent and criterion variables.

PREREQUISITES

1. Understand the concept of an operational definition.
2. Understand the different research purposes for investigating a problem.
3. Be able to state a research problem in a manner that permits investigation.

CONCEPTS

Each living creature, action, or situation that we experience in life has different characteristics, called variables, that can vary in amount, type, presence or absence, and single or multiple. Race car drivers have multiple patches on their driving suits to indicate their corporate sponsors. Each patch represents a variable of the type of corporate sponsor.

Each of us has many variables. Some of these variables, just like the patches on a race car driver's suit, are directly observable. What other variables are inferred and not directly observable?

How many different characteristics or variables do we have? Here are just a few: age, gender, eye color, economic status, religion, political affiliation, educational level, intelligence. How many different characteristics or variables are there in situations? Here are just a few: grading system of the instructor, tax rates, class size, neighborhood community, job classification, weather.

We go through our lives and try to make sense of events around us. Our anxiety is lessened if we can understand that events are not random and uncontrolled, but caused by someone or something. Those characteristics that we think cause something to happen are independent variables. Those characteristics that predict something are predictor variables. Do some people consider God to be the variable that causes what will happen?

An independent variable can cause a change or an effect in another variable. As we will later learn, the dependent variable is the characteristic that changes because of the independent variable. This rather simple description of the relationship between independent and dependent variables is fundamental to understanding the dynamics of research and to our actions. The assumption is that there is a cause for every effect.

Here is a mnemonic for remembering the difference between the independent and the dependent variables: ID. The independent variable must come first in time before the dependent variable to say that the independent variable caused the results in the dependent variable.

Just as Freud said that the id is the dynamic driving force for much of human behavior, the ID is the driving force for much research as we try to find causes (independent variables) for effects (dependent variables). (See, the tip to use "ID" to describe the relationship between independent and dependent variable shows that we really got something out of therapy for hearing all of those voices.)

The researcher directly or indirectly manipulates the independent variables of a potential cause of an effect. Direct manipulation is involved in quasi-experimental and experimental research questions with manipulation in any of the following ways:

Methods to Manipulate Independent Variables

- Use different amounts or degrees of a characteristic (e.g., three different amounts of a single medicine represent only one independent variable with three different amounts).
- Use different types or categories of a characteristic (e.g., two different methods to teach biology represent two different types of the single independent variable of method of teaching).
- Change the presence or absence of a characteristic (e.g., comparing persons with and without a college degree represents only the presence or absence of the single independent variable of college education).
- Simultaneously manipulate different numbers of independent variables (e.g., using different teaching methods combined with different amounts of instructional time represents two different independent variables: teaching method and instructional time).

A research study can simultaneously manipulate multiple independent variables. Simultaneously using multiple independent variables in a study is a factorial design. Using factorial or multiple independent variables in a study has the advantage of allowing examination of how the variables combine or interact to affect the results. This interaction is not available in studies in which only one independent variable is used. The factorial manipulation of multiple independent variables can involve different combinations of manipulations of type, amount, presence or absence, or the number of different independent variables.

A single factorial study with multiple independent variables has other advantages over multiple studies that each manipulate a single independent variable. A single factorial study can cut down on the time required to conduct multiple studies. A single factorial study may also reduce the procedural and implementation errors that are more likely to happen when conducting multiple studies.

Direct manipulation give us greater ability to assess the effects of the independent variables and is involved in quasi-experimental, experimental, single-subject, action, and mixed-methods research questions. However, people may already have preexisting differences on the independent variable. These preexisting differences create difficulties and limitations in determining the effects of the independent variable. Techniques to adjust for the preexisting differences among people are discussed in later chapters.

Indirect manipulation is involved in descriptive, comparative, causal-comparative, ex post facto, and ethnographic research questions.

Many errors in research are the result of falsely concluding direct manipulation of an independent variable. There are limitations in educational research studies comparing boys and girls. Indirect manipulation has less assurance of causality because of the inability to manipulate the assumed cause.

Does being a boy or a girl cause a difference in achievement, behavior, or other skills? Can different child-rearing practices, instructional practices, or other influences affect children's achievement and behavior? Can researchers directly manipulate the gender of a child? Can we perform sex-change operations in the interest of research? Does Sweden still perform sex-change operations? Did we not talk about ethics in a previous chapter?

Independent variables are called predictor variables when used to make predictions about the effects on another variable, which is called the criterion variable. The predictor variable(s) may also vary by (a) amounts, (b) type, (c) presence or absence, or (d) the number of predictor variables. In general, the use of multiple predictor variables in a study provides greater accuracy in prediction than the use of a single predictor variable.

Everyone tries to explain the world with assumptions of what variables can predict or cause other events to happen. Sometimes, we are just not aware of other variables that may cause or predict those events. This situation can be confusing, as it presents an endless range of possibilities to consider. How many times have we caught ourselves saying, "If I only knew then what I know now?"

Other variables beyond the specific independent or predictor variables investigated may affect and explain the results. The following section describes types of these variables that have potential influence upon results or decisions.

Categories of Extraneous and Other Variables Affecting Results

- Extraneous variables are characteristics or events that are not part of the study and provide an alternative explanation for the results for a specific group of people. There are many potential extraneous variables, such as the initial differences among people of different groups in important areas such as ability, education, and wealth. Many more of these extraneous variables are included in chapter 8, which explores threats to external and internal validity.
- Mediating variables try to explain how an independent variable affects the results. Mediating/intervening variables affect the relationship between independent and dependent variables separately from the effects of the independent variable. For example, practice can provide greater confidence to complete a task. Confidence is a mediating variable that contributes to the relationship between the independent variable of practice and the dependent variable of completion of a task.
- Moderating variables combine or interact with other independent variables. The effects of an independent variable are not consistent across all degrees or types of other independent variables. For example, there may be two different types of therapy for treating anxiety, but no overall difference between the effectiveness of the therapies. However, results show that the variable of the gender of the patients interacts with the type of therapy so that one type of therapy is more effective with females and the other type is more effective with males.
- Confounding variables are characteristics or variables that vary together in the same systematic way as with the other independent variables. Any separate potential effects of the confounding variable cannot be determined. Confounded variables need study as separate independent variables in order to determine the effects upon the results. Here are some examples: (a) Two different reading courses are used to teach reading to third-grade students, but one course is only taught by beginning teachers and the other is only taught by experienced teachers. It cannot be determined if the reading results are due to the type of reading course or the experience level of the teachers. (b) Three different types of ads are used to increase sales, but one type of ad is only on TV, another only on the

radio, and the third only in the newspaper. It cannot be determined if the sales results are due to the type of ad or to the outlet through which the ad is distributed.

- Multi-collinear variables are characteristics that vary simultaneously in the same way as the predictor variables when trying to answer a prediction question. The multi-collinear variable does not affect the accuracy of the prediction. For example, low socioeconomic status, poor diet, and lack of educational achievement are highly interrelated. This interrelation creates difficulty in determining the separate variables' role in predicting school dropout rates.

Researchers acknowledge this dilemma of extraneous variables by trying to consider and control for other variables that can explain the results. These variables can provide a different explanation; combine with other variables to affect results; or operate simultaneously to confuse the prediction or determination of the cause of events. Unfortunately, many of the conflicting results from research studies occur because of the various categories of uncontrolled extraneous variables as presented in this chapter.

Many characteristics or behaviors exist that we want to change or predict. The dependent variable is that characteristic or behavior that we investigate to determine if it is caused by the independent variable. The dependent variable is the effect, and the independent variable is the cause.

A research study can simultaneously have multiple dependent variables. A multivariate study is a study that simultaneously has multiple dependent variables. A multivariate investigation has an advantage in that one study can examine how people changed in multiple ways and provides a more comprehensive understanding of potential multiple effects caused by an independent variable. The multivariate study can have different combinations of types of dependent variables.

A single multivariate study has other advantages over multiple studies that each have a single dependent variable. The multivariate study is more efficient and can cut down on the time required to conduct multiple studies. As with factorial studies, a single multivariate study can reduce the implementation errors that may be inevitable when conducting multiple studies. If the question is not to determine the cause but to predict, the dependent variable is then called the criterion variable. To provide a more efficient and comprehensive understanding, there may be more than one criterion variable to be predicted.

The dependent and criterion variables must be operationally defined and accurately measured. The operational definitions of the dependent and criterion variables should accurately reflect the effect or criterion. The following are typical categories of dependent variables that are included in research studies.

Types of Dependent Variables

- time to complete a task
- quantity of a type of behavior
- accuracy of a behavior
- type or category of a behavior
- degree or intensity of a behavior
- percent correct of a behavior
- number of different behaviors
- frequency of different behaviors
- derived scores, such as averages and standard scores
- number of errors
- types of errors
- sequence of steps of a behavior
- quality of a behavior
- meeting a standard of behavior
- financial cost

The broad categories above reflect the different types of behaviors or measures that are part of our lives and the focus of research. Researchers develop specific measures that give information fitting into one of those broad categories. So much of what we do in our daily lives is to produce or contribute to changes in one of those broad categories. Those broad categories reflect the specific goals that we use to measure our lives, such as more money, better relations with a significant other, losing weight, better grades, getting a job, getting an education, increasing sales, and so many others.

Dependent or criterion variables of interest must be carefully selected. The use of multiple dependent or criterion variables allows investigation of multiple possible outcomes or results of interest. The old joke about someone saying that they have been happily married for five years but married for thirty years illustrates this point.

COMMON ERRORS

1. Not indicating the independent variable(s) manipulated
2. Incorrectly indicating direct or indirect manipulation of the independent variable(s)
3. Incorrectly indicating the number of amounts or degrees in the independent variable(s)
4. Incorrectly indicating the number of types in the independent variable(s)

5. Incorrectly indicating the presence or absence of an independent variable(s)
6. Incorrectly indicating the number of independent variables
7. Selecting an inappropriate range of degrees of the independent variables for the research question investigated
8. Selecting inappropriate intermediary degrees of the independent variables for the research question investigated
9. Selecting an inappropriate number of types of the independent variables for the research question investigated
10. Conducting multiple studies manipulating only a single independent variable rather than a single study manipulating multiple independent variables
11. Incorrectly indicating the degrees, types, or presence or absence of the predictor variable(s)
12. Incorrectly indicating the number of predictor variables
13. Including extraneous variables that affect the obtained results
14. Including mediating variables that intervene and affect the obtained results
15. Including moderating variables that interact with the independent variable(s) to affect the obtained results
16. Inaccurately attributing confounding variables to cause the obtained results
17. Including multi-collinear variables that are inaccurately attributed to the accuracy of the prediction
18. Using dependent variables without an accurate operational definition
19. Using dependent variables that do not have a range of amounts or categories to measure the full range of effects
20. Using dependent variables that do not accurately represent the characteristic effect of interest
21. Using dependent variables that do not measure latent or delayed effects
22. Using an insufficient number of dependent variables for the characteristic investigated
23. Conducting multiple studies with a single dependent variable rather than a single study with multiple dependent variables
24. Using a criterion variable that does not have an accurate operational definition
25. Using a criterion variable that does not accurately represent the predicted characteristic

EXAMPLES

- *The Wall Street Journal* reported that Johnson & Johnson did not adequately test a hip implant device used in more than 90,000 patients. Laboratory equipment tested only one angle of replacement. Later results indicated that the implant failed at different degrees or amounts of angles of implant. The error here is not exploring an appropriate range of degrees of the independent variable.
- A painting by Monet presents a different image based upon the independent variable of viewing distance from the painting. A close viewing provides an image of disorganized strokes, but greater distance provides a cohesive image. An error occurs from not examining the effects from different amounts of the independent variable of distance from the painting.
- A common occurrence in daily life is that it is not always "what" we say, but the additional independent variable of "how" we say something. For example, the phrase "Yeah, right." Does the additional independent variable of how we said that phrase make a difference? (We just discovered a new mathematical principle whereby the two positives of "yeah" and "right" make a negative.)
- *The Plain Dealer* reported a study by the Northern California Cancer Center that showed that exposure to sunlight reduces the risk of breast cancer. The study did not report how the needed exposure was defined in terms of both the amount of time and the amount of the body exposed to sunlight. Could errors occur from different amounts of the independent variable of sunlight? Is omitting an important additional dependent variable, the risk of skin cancer, an error of practical importance? (Any volunteers to go to the French Riviera to conduct research? Hopefully there is a grant agency that would provide funds for this research.)
- *USA Today* reported that James Garbarino, a psychologist at Cornell University, reviewed the literature and found that it is erroneous to conclude that any one risk factor, such as violent media, predicts youth violence. There are many uncontrolled and potentially multi-collinear factors, including feeling socially isolated, being bullied, mood problems, and family situation. Looking for one factor may be an error of not including a sufficient number of independent and predictor variables.
- *The Plain Dealer* summarized a report from the Federal Institute of Technology in Zurich stating that glaciers in the Alps melted faster in the cooler 1940s than now. It is an error to conclude that there is no global warming. The reason for the obtained result is the less polluted atmosphere during the 1940s, mediated by allowing more solar radiation to reach and melt the glaciers than the more polluted atmosphere of today. An error is not considering the potential mediating variable of pollution.

- Examine the label on prescription medicines. Instructions often prescribe taking the medicine with food or water, but not with other medicines or alcohol. These interactions combine and moderate the effect of the prescription medicines. Just keep a record of the number of personalities we see in the news who die of drug overdose or taking drugs and other substances that interact or combine. These are errors of ignoring the potential interaction with moderating variables.
- *The New York Times* reported that researchers at the University of Turku, in Finland, found that persons without pets could be predicted to be healthier than pet owners. Pet owners in Finland tend to be older, poorer, and less educated, and have to take pets out into the cold of the Finnish winter. These additional variables are multi-collinear with pet ownership. The error is not considering these multi-collinear variables as additional predictors of illness. Are these cold, hard facts in research?
- *The Plain Dealer* reported that Margaret Cicirella of Case Western Reserve University School of Medicine found that just lowering the amount of the independent variable of salt intake is insufficient to prevent hypertension. An additional moderating variable of a diet high in potassium may moderate the effects of salt intake and lower the incidence of hypertension. An error is in not considering the additional moderating variable of potassium intake.
- *The New York Times* reported that many government agencies determine the "value of a statistical life" in calculating a cost-benefit analysis of governmental regulations. Different agencies may place a different value on life. The Bush administration cut the value of a human life by over $1 million to lessen the benefit and weaken the need for more regulations from the Environmental Protection Agency. How do we operationally define the dependent variable of "human life"? (By the way, how do we also measure the "sanctity of life"?)
- *The Plain Dealer* reported that Pfizer, Merck, and Roche Holding AG have spent over $1 billion to find drugs that raise the dependent variable of HDL cholesterol because of the assumed benefit in lowering the risk of heart attack. Copenhagen University Hospital conducted research indicating that HDL may not be the most important dependent variable. The error may be that another dependent variable, high levels of triglycerides, may be a more important dependent variable on which to focus to reduce heart attacks.
- Here is a personal question. What is the most important dependent variable that indicates success in college? Is it the final grade point average, whether we get a job upon graduation, whether we find a significant other, or how fast we can chug a six-pack? Determining the significant dependent variables or results in different situations is an important task for all of us. What different variables should be in a rating system? Do errors

occur in selecting a university to attend because of using inappropriate variables?
- Income inequality is an important societal issue that has even been mentioned by Pope Francis. Different research studies have operationally defined income and income inequality in different ways—for example: before or after taxes, before or after tax credits, before or after government transfers like Social Security, before or after food stamps, and other considerations. What is included in the operational definition of the dependent variable of income does make a difference and can contribute to errors in conclusions and policies.

REDUCING ERRORS

Selection of appropriate variables can help to reduce errors. The variables should represent those characteristics of importance that are included in the problem and the hypothesis. Each variable must have an operational definition that allows accurate measurement.

Researchers are required to indicate if the independent variable was directly or indirectly manipulated. Indirect manipulation is more limited in making causal conclusions. Erroneously stating a direct manipulation of an independent variable creates significant misinterpretation of the results because of the illusion of control of potential causes.

A sufficient number and range of categories or scores in the dependent measures are required. Having too few measures or measures with limited range can contribute to an inaccurate decision by not identifying potential effects. Including more measures and more types of measures of the dependent variable gives greater confidence in the comprehensiveness of the results and the accuracy of the decision.

Due diligence must also be undertaken for other types of extraneous variables. These variables can potentially influence the outcomes and conclusions. A frequent reason for conflicting results from different research studies is ignoring potential other types of extraneous variables.

BEST PRACTICES

- Accurately indicate how the independent or predictor variables vary (type, amount, presence or absence, and single or multiple).
- Indicate if the independent variables is directly or indirectly manipulated.
- Specify the potential influence of other variables (extraneous, mediating, moderating, confounding, and multi-collinear).
- Specify the type of information in the dependent or criterion variables (time, quantity, accuracy, type or category, degree, percent correct, num-

ber of different behaviors, frequency of different behaviors, derived scores, number of errors, types of errors, sequence of steps, quality of behavior, meeting a standard, financial cost).

- Select dependent variables with sufficient sensitivity and range to detect meaningful and extreme changes.

INTERRELATION TO OTHER TOPICS

This chapter presented the concepts of the independent (predictor), dependent (criterion), and extraneous variables. Just as DNA gives the genetic blueprint for life, the different types of variables give life to the research study in gathering information. Extraneous variables exist that can alter the determination of causality or prediction, just as do genetic mutations.

Suggestions were included for procedures to minimize the errors or mutations in configuring the arrangement of variables. Examples were included of other factors in everyday life that may contribute to erroneous decisions.

The next chapter provides important concepts and procedures to measure the obtained information. As we will see, an inappropriate level of measurement or a variety of other measurement considerations can contribute to errors.

Chapter Five

Measurement Foundations

"How Big Is Infinity?"

The shifts in Fortune test the reliability of friends.

—Cicero

Good data are the weapons of myth destruction.

—Anonymous

Our ideas are only intellectual instruments which we use to break into phenomena.

—Claude Bernard

The accuracy of the report being always tried . . . but I shall be content if it is judged useful by those inquirers who desire an exact knowledge.

—Thucydides

SIGNIFICANCE

Gathering information requires that we have precision in determining the type, rank, or degree of a characteristic that an individual or a group possesses. Measurement means determining the smallest unit of difference observed on that characteristic. The result is the determination of the classification or magnitude of a characteristic in terms of a unit of measurement.

These data must appropriately measure the characteristic that we intend (validity) and provide a consistency from one set of measurements to another (reliability). The previous chapter identified specific types of dependent variables. Operational definition of those variables requires measures that are valid and reliable. This chapter will present ways that both validity and reliability are determined. Data that have weak validity and reliability can contribute to erroneous decisions.

Norm-referenced measurement measures how an individual compares to others in a group or how a group compares to another group. A norm group is the group used as the basis for comparison and has an established standard of achievement or behavior. Comparison to an inappropriate norm group can distort the evidence and contribute to erroneous conclusions.

Data collected regarding how an individual or group compares to a standard of achievement is criterion-referenced measurement. Standards that are set too high or too low can also distort the evidence and contribute to erroneous conclusions.

Formative assessment entails collecting data during the process of a study or activity rather than developing a summative assessment at the end of an activity. Many errors come from only collecting data at the end of a study, which prevents changing or modifying ineffective practices during the study.

Data can have multiple dimensions or characteristics. Measuring those characteristics requires establishing an operational definition of each. A rubric indicates the degree to which people differ on that characteristic. Rubrics that are unreliable, qualitative, vague, and/or have too small or too large a difference on the characteristic contribute to errors in making decisions.

Errors can also reflect the use of a holistic scoring method that combines multiple dimensions into one overall summary of performance. Not all combinations of the categories of the different dimensions may be possible, so that holistic scoring has errors because of unreliability. Analytic scoring provides a separate score for each dimension, and this increases the possibility of reliability.

The level of precision in measuring a characteristic has an influence upon the interpretation of information. Significant errors in decision making can be the result of imprecise levels of measurement of a characteristic. Less precise levels of measurement can have lower reliability because of the variability in interpretation.

All of these concerns can contribute to errors and erroneous decisions. Understanding the potential errors can assist in making more accurate interpretations and decisions.

LEARNING OBJECTIVES

1. Understand the concept of validity.
2. Understand the different ways to demonstrate validity:

a. Construct validity
b. Content validity
c. Criterion-related validity

3. Understand the concept of reliability.
4. Understand the different ways to demonstrate reliability:

a. Test-retest or stability
b. Parallel or equivalent forms
c. Internal consistency or split half
d. Rater or scorer consistency

5. Understand the concept of standard of measurement.
6. Demonstrate ability to interpret a standard error of measurement.
7. Understand norm-referenced assessment and interpretation.
8. Understand the selection of an appropriate norm group.
9. Understand criterion-referenced assessment.
10. Understand the establishment of appropriate criterion standards.
11. Understand formative assessment.
12. Understand summative assessment.
13. Understand the difference between scoring dimension and standards and rubrics.
14. Understand methods to construct a rubric.
15. Understand the advantage of analytic scoring compared to holistic scoring.
16. Understand the different levels of measurement (nominal, ordinal, interval, and ratio).

PREREQUISITES

1. Understand the concept of dependent variable.
2. Be able to apply the concept of operational definition.

CONCEPTS

Many terms that we have used in our everyday conversation can also help to identify appropriate research instruments or ways to collect accurate information. For example, we talk about information that is consistent (reliable), measures the characteristics that we want (valid), tells us whether the person does more or less what everyone else does (normal), or whether a person does better or worse than a standard (criterion). We also collect information during the process of learning or performing a task (formative) or after we have completed a task (summative).

Researchers give specific meanings to those terms in parentheses, but we do not need a secret decoder ring or a closed-caption interpreter for translating these everyday terms into more precise research terminology. Researchers need to use instruments to collect information that have high amounts of both validity and reliability.

Mathematics can calculate the degree of validity and reliability, but we will leave the mathematical ways for other textbooks to present. Can we hear one loud cheer for that?

For those afraid of mathematics, close your eyes and avoid the next sentence. This sentence is for those who need a mathematics "fix," as appropriate levels of both validity and reliability should be at or above .80 with a maximum value of 1.00. For the rest of us, we can now open our eyes and keep reading with the comforting thought that we just need to know that different instruments to collect data should provide both valid and reliable information.

The next chapter describes different types of data collection instruments. To be valid, the instrument must provide evidence in each of the following three ways:

Methods of Estimating Validity

- Construct validity: The instrument measures the intended characteristic or variable, such as intelligence, reading achievement, or employee morale. The instrument used provides the operational definition of that characteristic.
- Content validity: The instrument measures enough of the characteristic or construct. Three-minute speed dating is not enough time to get a valid measure of another person's compatibility.
- Criterion-related validity: The instrument is valid if it provides information that demonstrates a relationship to the criterion or standard of other instruments that measured the construct. The data from an instrument should be related to other, similar instruments or characteristics and not related to dissimilar instruments or characteristics.

A measure that has one type of validity may not have another type of validity. For example, a measure may have construct validity, but not content validity. A valid measure requires demonstration of each of the types of validity.

Reliability is another important foundational concept in determining the quality of a research instrument. We have greater confidence in the information we collect if we know that instrument will provide consistent results. To be reliable, an instrument must provide evidence in one of four following ways:

Methods of Estimating Reliability

- Test-retest or stability indicates the degree of consistency of a person over time on the same instrument.
- Parallel or equivalent forms indicate the degree of consistency of a person on different versions or forms of the instrument at the same or a later time.
- Internal consistency or split-half indicates the consistency of a person throughout the assessment, such as the beginning to the end, or odd-number items compared to even-number items.
- Inter-rater or inter-scorer indicates the degree of consistency between different raters or scorers in judging a behavior, product, or performance.

A measure that has one type of reliability may not have another type of reliability. For example, a measure may have internal consistency reliability, but not parallel forms reliability. A reliable measure requires demonstration of the type of reliability appropriate for the decision.

Once the degree of reliability has been determined, a standard error of measurement (SEM) is calculated. The SEM shows the range above and below the person's score expected by that person on retesting, on a different form, or between different raters or scorers. As the reliability increases, the standard error of measurement decreases, and the confidence in the consistency of the person's score rises. The standard error of measurement is an important guide to interpreting a typical score for a person. Many errors falsely assume a score as typical for a person rather than a range of possible scores.

In addition to concerns about validity and reliability, we must determine if we are comparing a person to others in a comparison group called the "norm group" (norm-referenced) or to a standard (criterion-referenced). There are many possible norm groups. It is important that the norm group used as a comparison be appropriate. Errors result from comparison to inappropriate norm groups.

An individual can have a different relative position or standing in different groups. In a norm-referenced assessment, it is irrelevant how well a person performs a task. It is only how well a person does in comparison to others in that norm group. Medical evaluations often report how an individual compares to others of that age. A 90-year-old individual could be in good health compared to other 90-year-olds, but not compared to 20-year-old individuals.

In criterion-referenced assessment, the person compares to a standard or a criterion on any of a variety of measures, e.g., completion, speed, accuracy, length, time, money saved. It is only how well a person performed the task, and how others performed is irrelevant. The criterion must be set at an appropriate level for any of these or other measures.

Do we set standards for who will be our friends and significant others? If the standard is set too low, then many will meet the standard. If the standard is set too high, then few will meet the standard, and we will have few friends. Do we make judgments about ourselves by comparison to others? Do we make judgments about ourselves compared to a standard? Which comparison is better? Do we make ourselves feel better by making comparisons to the less fortunate?

An additional consideration in the collection of data by one of the research instruments is whether the collection occurs during the process (formative) or at the end of the process (summative). The advantage of formative data collection is that it provides the opportunity to change or modify treatments during the study (think of a football team changing tactics at half time). The advantage of summative data collection is that it provides the opportunity to collect data at the conclusion of the process to certify achievement of a product or result.

The data reflect the operational definition of the dependent variables of the study and can include a variety of scoring standards or dimensions. It is important to be precise regarding the particular dimensions or scoring standards in the obtained data.

Rubrics provide the method by which to indicate the degree to which people differ on each scoring standard or dimension. The rubrics should permit a reliable differentiation between meaningful degrees of achievement on the scoring standard. The use of more quantitative methods of counting or percentages provides higher levels of reliability compared to the use of more qualitative and vague terms such as "excellent" or "good."

Sometimes the rubric for differentiating among levels on the scoring standard is either too large or too small. A major league baseball player making a hit 30% percent of the time, or with a lifetime batting average of .300, will be in the Hall of Fame. A major league baseball player making a hit 28% of the time, or with a lifetime batting average of .280, will not likely be in the Hall of Fame. This rubric to differentiate Hall of Fame players may be of limited practical significance, as there is a difference between the two players of only 10 hits in 500 times at bat. However, the lifetime earnings may differ by millions.

Analytic scoring for each separate scoring standard generally provides more reliable information than an overall or holistic score from all scoring standards. Holistic scoring may not have all combinations of categories of multiple standards and, therefore, is less reliable.

The collected data also have four different scales of measurement precision. The level of measurement can affect the selection of appropriate statistical analysis and the interpretation of the data. The following section describes the four levels of measurement.

Levels of Measurement

1. Nominal data only indicate whether two or more persons are the same or in different categories on such characteristics as gender or religious affiliation.
2. Ordinal data indicate a rank order, but the ranking does not indicate how much of a difference there is between persons ranked first, second, and so on.
3. Interval data indicate how much of a difference exists between persons with different scores because there is an equal interval of that characteristic between each score.
4. Ratio data have the precision of interval data, but also an absolute zero as a possible score.

Below is a hypothetical illustration of how level of measurement affects the interpretation and the decision. For each comparison, please select the box of cereal that has the most sugar:

- Comparison 1:
    a. Super Sugar Slices
    b. Organic Goodness Flakes

- Comparison 2:
    a. Super Sugar Slices (ingredients in order: oats, sugar, corn, wheat)
    b. Organic Goodness Flakes (ingredients in order: sugar, corn, oats, wheat)

- Comparison 3:
    a. Super Sugar Slices (ingredients: oats 48%, sugar 47%, corn 3%, wheat 2%)
    b. Organic Goodness Flakes (ingredients: sugar 26%, corn 25%, oats 25%, wheat 24%)

The increased level of measurement contributed to a different interpretation of which box had the most sugar. Did you see how the selection of the cereal box with the most sugar changed as the level of measurement precision changed?

This is a common occurrence in daily life, as arguments often result from differing interpretations of a less precise measurement. We sometimes have a different understanding than other people of a vague term or label such as "bureaucracy," "liberal," "conservative," "justice," and many others. The level of measurement is a concern in identifying errors in interpreting research information and even errors in communicating with each other.

COMMON ERRORS

1. Construct validity weak or not reported
2. Content validity weak or not reported
3. Criterion-related validity weak or not reported
4. Test-retest or stability reliability weak or not reported
5. Parallel or equivalent form reliability weak or not reported
6. Internal consistency or split-half reliability weak or not reported
7. Inter-rater or inter-scorer reliability weak or not reported
8. Standard error of measurement large or not reported
9. Standard error of measurement not interpreted in a decision
10. Inappropriate norm group selected for comparison
11. Inappropriate standard for criterion-referenced assessment where standard is too high or too low
12. Inappropriate comparison using norm-referenced rather than criterion-referenced assessment
13. Inappropriate comparison using criterion-referenced rather than norm-referenced assessment
14. Inaccurate consideration of formative assessment data as summative assessment
15. Inaccurate consideration of summative assessment data as formative assessment
16. Analytic scoring of each standard not used
17. Scoring criteria not accurately operationally defined
18. Use of scoring standards with multiple dimensions or criteria in which not all factorial combinations of degrees or amounts are possible
19. Incorporation of rubrics for scoring that use qualitative terms such as "minimal" or "excellent" without precise operational definition
20. Use of rubrics with "gaps" (e.g., 0–5) in which any score in the gap can be recorded without operational definition
21. Rubrics truncated so unable to differentiate the full range of potential responses
22. Rubrics not accurately operationally defined
23. Inaccurate level of measurement assumed
24. Level of measurement affects conclusion

EXAMPLES

- *The Plain Dealer* reported that the Intoxilyzer 8000 was an instrument certified by the Ohio Department of Health for use in Ohio to test suspects for drunk driving. The instrument had low reliability because it was subject to errors based upon how long the suspect breathed into the instrument and the environmental temperature. Do we think that the low reliability could make the results vulnerable to legal challenge? No, the Ohio Supreme Court decided that defendants could not challenge the reliability of breath-testing instruments after certification by the Ohio Department of Health. How many erroneous convictions occurred because of an unreliable instrument?
- A criterion standard for accuracy is often set at an assumed high percentage, but that standard may still be erroneously too low to reflect the practical significance of not meeting the criterion. Having a 99.9% accuracy criterion results in 20,000 incorrect drug prescriptions filled each year and 16,000 pieces of mail lost by the United States Postal Service every hour. Are criterion standards too high or too low? What should the criterion standard be in a research study? Are there errors because standards are too low or too high?
- *The Plain Dealer* reported that the National Commission on Forensic Science is establishing national forensic science standards and practices for crime labs. No current standard practices exist, and there is significant variability or unreliability in the frequency of errors made by crime labs in obtaining and testifying about results. No one knows how often errors are made in the evidence presented. Faulty forensic evidence was second only to erroneous eyewitness testimony in erroneous convictions. Can we be certain beyond a reasonable doubt of the accuracy or errors in the evidence? What is reasonable? What is a doubt?
- Do we make personal errors in judging how well we are doing by comparing ourselves to a norm group rather than to an internal criterion? How much of our happiness have we sacrificed in pursuit of trying to keep up with the "norm" group of the images of young and beautiful people shown by advertisers on TV? Is this an error of using a norm-referenced interpretation of how we are doing, rather than a criterion-referenced interpretation according to our own standard?
- History has provided many examples of how increasingly precise measurement has identified errors in knowledge and contributed to a more complete understanding. Discoveries of ways to measure the speed of light, the distance of galactic objects, and cellular processes and atomic particles have all provided specific information that corrected erroneous understanding.

- *USA Today* reported that different states have different IQ standards for identification of intellectual disabilities in determining imposition of the death penalty for crimes. Some states use an IQ below 70 and some use an IQ between 70 and 80. Freddie Hall was originally tested and obtained an IQ of 60, but two later tests determined his IQ was between 70 and 80. The degree of reliability of a test contributes to errors in judging intellectual disability. Are there life-and-death consequences of errors resulting from the unreliability of measures? This case is now before the Supreme Court.
- Driving at the speed limit uses a criterion-referenced measurement. Driving above the speed limit but with the flow of traffic is a norm-referenced measurement. Is it an error to drive according to norm-referenced measurement standards rather than criterion-referenced measurement standards? Do we think the officer will understand when we explain that we just confused which measurement system to use in driving speed? Do we think that officers will accept the excuse, "I was only driving by a norm-referenced assessment"?

REDUCING ERRORS

Selecting measuring instruments that have appropriate levels of validity and reliability reduces errors. Validity gives evidence of accurately measuring the intended behavior. Reliability gives evidence of the consistency of measurement. A smaller standard error of measurement indicates greater reliability and can contribute to a reduced probability of errors.

Errors occur in comparing information to an inappropriate norm group in norm-referenced measurement or in establishing criterion standards that are too low or too high. Comparison of the results from an individual or group to a representative norm group can prevent errors in norm-referenced measurement. Comparison of criterion-referenced standards to professional standards or to other research can reduce errors in establishing appropriate criterion standards. Examination of the comparison group or the specific criterion standards can prevent errors of misinterpretation.

Many different types of data are available, including inconsistencies between these different forms of data that can contribute to errors in interpretation. Use of multiple types of measures can reduce these errors by providing a wider and more representative indication of the performance or behavior of an individual or group.

Formative assessment reduces errors by detecting ineffective or inaccurate practices during a study. This early detection provides an opportunity to discontinue erroneous practices.

Inappropriate scoring standards and rubrics can create errors in terms of failing to detect or exaggerating differences in levels of performance or behavior. Prevention of these errors requires clear operational definitions of the scoring standards or criteria. These standards must be directly observable.

A rubric indicates the different degrees or types of behaviors measured on a standard or criterion. Any of the following strategies can reduce errors: the use of quantitative rather than qualitative terms, analytic scoring in which each scoring standard has one dimension, the use of rubrics with a greater range of potential performance, and the use of rubrics with sensitivity to detect meaningful differences in behavior.

An accurate and precise level of measurement can also reduce errors. The level of measurement is both an important consideration in the selection of the appropriate method of statistical analysis and a limit to the level of precision in interpretation of data and the decision. Inaccurate and/or imprecise levels of measurement contribute to errors in selecting appropriate statistical analyses and to the level of precision in a conclusion.

BEST PRACTICES

- Evidence for validity (construct, content, criterion-related) should be r = .80 and above.
- Evidence for reliability (test-retest, parallel forms, internal consistency, and inter-rater) should be r = .80 and above.
- Use a small standard error of measurement for individual scores (4% variation or less from obtained score).
- Norm-referenced assessment should have an appropriate comparison norm group.
- Use a level of criterion in criterion-referenced assessment that accurately reflects the importance of consequences of not attainting the standard (high standard for significant consequence).
- Provide an observable and clearly defined separate score for each standard or dimension.
- Use rubrics that permit reliable differentiation of important degrees or categories of performance (avoid vague or qualitative terms).
- Use interval or ratio level of measurement.

INTERRELATION TO OTHER TOPICS

Regardless of the type of data collection instrument, each instrument must be valid and reliable and provide a sufficiently precise level of measurement. The foundations of measurement apply to each type of instrument.

The type of data collection instrument indicates the operational definition of the dependent variable. The data collection instruments used should be capable of providing evidence to support or refute the research hypothesis. Each type of data collection instrument has advantages and disadvantages. Comparison of these different instruments will be included in the next chapter.

Chapter Six

Data Collection Instruments

"How Do I Measure Thee? Let Me Count the Ways"

It is better to know some of the questions than all of the answers.

—James Thurber

Because I have answered three questions, and that is enough.

—Lewis Carroll

Guess if you can, choose if you dare.

—Pierre Corneille

It is not every question that deserves an answer.

—Publilius Syrus

Ask me no questions, and I'll tell you no fibs.

—Oliver Goldsmith

Undoubtedly we have no questions to ask which are unanswerable.

—Ralph Waldo Emerson

We have ways of making men talk.

—Waldemar Young

The Socratic manner is not a game at which two can play. Please answer my question, to the best of your ability.

—Sir Max Beerbohm

You can observe a lot by watching.

—Yogi Berra

In the fields of observation, chance favors only the prepared mind.

—Louis Pasteur

You see, but you do not observe. The distinction is clear.

—Sir Arthur Conan Doyle

Faith is the substance of things hoped for, the evidence of things not seen.

—The Epistle of Paul the Apostle to the Hebrews

Saying is one thing and doing is another.

—Michel de Montaigne

Since the measuring device has been constructed by the observer . . . we have to remember that what we observe is not nature in itself but nature exposed to our method.

—Werner Heisenberg

Judge not according to the appearance.

—The Holy Bible

The historian, essentially, wants more documents than he can really use.

—Henry James

Every intellectual product must be judged from the point of view of the age and the people in which it was produced.

—Walter Pater

Writers . . . frequently fall into a habit which can neither be cured nor pardoned—the habit of making history into the proof of their theories.

—John Acton

Anybody can make history. Only a great man can write it.

—Oscar Wilde

Let's look at the record.

—Alfred Smith

Art! Who comprehends her? With whom can one consult concerning this great goddess?

—Beethoven

If a poet interprets a poem of his own he limits its suggestibility.

—W. B. Yeats

SIGNIFICANCE

Examining the specific data that are collected is a significant procedure in determining the accuracy of evidence. Too often, errors result from inaccurate data or an inappropriate data collection instrument.

Previous chapters indicated that issues of inappropriate levels of validity and reliability lead to erroneous information. Less precise levels of measurement and scoring procedures with rubrics that do not detect sufficient degrees or amounts of difference on a standard can also contribute to errors.

Errors can occur in a variety of stages in the development and application of the data collection instrument. An error in any stage can have significant adverse consequences for the accuracy of the obtained information. These stages include the following:

- construction of the instrument
- procedures in administering the instrument
- recording and collecting the data
- verifying appropriateness of the instrument for the question to be answered
- interpretation of the data

This chapter examines advantages and limitations of each specific type of data collection instrument and identifies errors that can occur with each type. The following are the five broad types of data collection instruments:

1. Tests
2. Questionnaires
3. Interviews
4. Observations
5. Artifacts and records

TESTS

Learning Objectives

1. Understand the following different types of tests:

    a. Intelligence (aptitude)
    b. Achievement

c. Vocational (career)
d. Personality (objective)
e. Personality (projective)

2. Understand the following formats for test items:

    a. Completion
    b. True-false (binary choice)
    c. Matching
    d. Multiple-choice
    e. Essay

3. Understand that test takers differentially prepare for a test based upon the item format.
4. Understand differential implications of penalties for guessing.
5. Understand differential implications of blank or no-response answers.
6. Understand differential implications of tests administered individually or in groups.
7. Understand a prerequisite for sufficient training of test administrators.
8. Understand differential implications of tests administered with a time limit (speed) compared to no time limit (power).
9. Understand that a lengthy test may have differential implications regarding the fatigue of the test taker.
10. Understand that the question wording and structure may differentially affect the responses.
11. Understand cultural influences that differentially affect the responses.
12. Understand characteristics of the test administrator that affect responses during individually administered tests.
13. Understand that the anxiety of the test taker may affect responses.
14. Understand that tests provide only an estimate of a person's intelligence, achievement, vocational preferences, and personality, but do not indicate the worth of a person.

Prerequisite

1. Understand measurement foundations of data collection instruments.

Concepts

We have taken many tests in our life, and we did well on some and not so well on others. Perhaps we did not do as well because we did not prepare, or because of the format of the test or the wording of the test items. Were the tests that we took confusing, hard, easy, or did we just not understand?

There are different types of tests. The first is intelligence (aptitude) tests that measure the potential of a person for learning and typically reports an IQ score. The IQ score indicates a norm-referenced comparison to others in the norm group. An IQ of 100 is average for the norm group. An important concept to remember is that a person is more than just an IQ score. Success in a variety of situations may also depend upon motivation, personality, and experience.

The second type is achievement tests that measure the degree to which a person has acquired an understanding of the content of academic, technical, or performance skills. Norm-referenced assessment makes a comparison to the achievement of others in the norm group. Criterion-referenced assessment makes a comparison to a standard of achievement, such as the percent correct, time to complete, or mastery of a level of accuracy. Tests may also provide diagnostic information such as types of misunderstandings.

The third type is vocational (career) tests that measure a person's career interests and skills for different careers. This information can help to guide the selection of a career.

The fourth and fifth types are personality tests that measure the underlying personal perceptions and dynamics that influence our attitudes and behaviors. These tests may be objective, in which the person responds by selecting an alternative answer, or projective, in which the person supplies an answer to an ambiguous photo or word.

The formats or the ways in which items are constructed affect how we study and respond. There are five common types of item formats:

Types of Test Item Formats

1. Completion items have a blank in a sentence, and the person has to supply the correct answer that completes the sentence.
2. True-or-false or binary choice items have the person select whether a statement is true or false.
3. Matching items have a set of premises and responses, and the person has to select or match one of the responses to a premise. Care must be taken to minimize the chance that a person will simply use process of elimination; the instructions should indicate that a response can be used once, more than once, or not at all.
4. Multiple-choice items require the person to select or choose an alternative term or statement that best addresses the question.
5. Essay items require the person to construct an answer that can have varying length and complexity of ideas. There must be clear scoring standards and a rubric for responses to essay test items.

An individual or group of persons can be administered a test. It is important that test administrators receive appropriate training to ensure uniformity in administration of the test. Persons may respond differently to tests administered individually versus in groups and may affect the results of a research study.

It is common for some people to become anxious in taking a test, and precautions should be taken to minimize the effects of anxiety on test performance. Anxiety can increase because of (a) characteristics of the test administrator, (b) time limits for completing the test, (c) a penalty for a blank answer, and (d) a penalty for an incorrect answer. The penalty for guessing differentially affects responses based upon personality and makes some of us hesitant to answer unless we believe that we are certain of the answer. Do we get nervous when we take a test?

Other concerns that affect the responses to test items include (a) wording of the test item, (b) fatigue from a lengthy test, and (c) cultural background of the test takers. Each test item needs review to determine the potential effect on the responses.

Although the specific test used provides an operational definition of the characteristic that is measured, it is important to remember that the results of any test provide only one indicator of a person's ability, achievement, interests, and personality. Many other considerations affect the usefulness of any test result in making decisions.

Common Errors

- For all types of test item formats:
    1. Test administrators do not receive appropriate training to correctly administer the test.
    2. Persons do not receive appropriate training to record correctly and/or score responses.
    3. Responses are recorded erroneously on answer sheets.
    4. A long testing session or large number of test items introduces fatigue.
    5. The amount of time allotted to complete the test is too short.
    6. The test item format is inappropriate for the concept or topic.
    7. Characteristics of the test administrator affect responses during individual test administration.
    8. A penalty for incorrect or blank answers differentially affects response rate.
    9. Characteristics of the testing environment affect responses during group test administration.

10. Order of test questions provides additional clues or affects responses to subsequent questions.
11. Negative phrasing is used in questions.
12. A cultural or stereotype bias exists in items.

- Additional errors for completion items:

13. Grammatical clues to correct answers are included.
14. Precision of answers is not indicated.
15. Completion blank is not placed at the end of a statement.
16. Multiple completion blanks exist in a statement.
17. The length of completion blank provides a clue to correct answers.
18. Multiple possible answers exist for a blank.

- Additional errors for true-false or binary-choice items:

19. Multiple ideas are included in the question.
20. Unequivocal terms are used in the question, such as "always" or "never."
21. Vague terms are used in the question, such as "sometimes," "occasionally," and "frequently."

- Additional errors for matching items:

22. Responses are used only once, enabling process of elimination.
23. The test does not state that responses should be used once, more than once, or not at all.
24. Possible responses are not homogeneous in content or categories.
25. Premises are not homogeneous in content or categories.
26. Possible responses are not logically ordered.
27. Premises are not logically ordered.
28. Grammatical clues are included for the correct match.
29. There is a lack of clarity in how subjects record answers, which can contribute to inaccuracy.

- Additional errors for multiple-choice items:

30. Compound alternatives are used as the correct choice, such as A and C or B, C, and D.
31. Terms implying a degree of correctness, such as "most correct," are included.
32. Possible choices are not logically plausible.
33. Possible choices are not homogeneous regarding content or categories.
34. Possible choices do not have parallel length.
35. Possible choices are not in logical order, e.g., alphabetical, chronological, or organized by amount, distance, or size.

36. Possible choices do not employ parallel grammatical complexity.

- Additional errors for essay items:

37. Respondent does not know amount of time, length of answer, or points for the answer.
38. The absolute amount of detail to include in the answer is either not stated or vaguely stated, in terms such as "at least" or "minimum."
39. A choice of topics is allowed, which introduces potential variation among test takers.
40. There is a lack of clarity regarding how the responses should be structured or which parts of the test item to answer.
41. The role of writing mechanics in grading answers is not clearly stated.
42. Scoring is not anonymous.
43. Answers for all essay questions for the same person are scored before scoring the answers of other persons.

Examples

- *USA Today* reported that the National Merit Scholarship program awards scholarships using different test standards on the Preliminary Scholastic Aptitude Test (PSAT) for students in different states. The standard selects from the top fraction of a percentage of test takers in that state. The lack of a uniform standard means each student has a different norm comparison group to determine eligibility to receive a scholarship. For example, students in Arkansas with lower PSAT scores than students in Connecticut receive a National Merit scholarship. Are there errors in college admission because of different National Merit Scholar standards? Are there ethical errors based upon differential selection?
- *The New York Times* reported that scoring errors made by test publisher CTB/McGraw-Hill on the TerraNova exam resulted in 8,600 test takers in New York City public schools incorrectly needing to go to summer school. Extra costs and aggravation were the result of this error.
- MSN.com reported that errors in interpreting test results reflect different operational definitions of a characteristic and different item formats. For example, the 12-item Life Orientation Test defines optimism as dispositional, but the Content Analysis of Verbatim Explanations defined it as an explanatory style. Each of these tests represents a different operational definition of the same characteristic, optimism, but with different results. Errors in interpretation may result regarding which test is used to measure optimism.
- A classic example of the limitations of the use of test scores is the "Larry P." case, in which the IQ test had cultural biases that contributed to the

identification of a higher proportion of minority children as "mentally retarded." A child can no longer be identified as "mentally retarded" (the correct phrase is now "intellectually disabled") based upon a single test score, and the tests used should also minimize cultural influences.

QUESTIONNAIRES

Learning Objectives

1. Understand that question wording may affect the responses.
2. Understand that contextual factors affect responses. These factors include anonymity, sponsorship, opportunity to clarify questions, length of questionnaire and type of response permitted, social desirability, order of questions, and administration by mail rather than in person.
3. Understand that the response rate to the questionnaire may affect the representativeness of the results.
4. Understand that the response format (e.g., forced-choice, paired comparisons, Likert scale, numerical rating, ranking) may affect the response and the interpretation of the response.
5. Understand that questionnaires represent only one way to collect data.

Prerequisite

1. Understand measurement foundations of data collection instruments.

Concepts

Questionnaires are among the most commonly used types of instruments to collect data. We have all completed many questionnaires designed to obtain our opinions or beliefs. We may not have completed or returned every questionnaire, as some seemed too long, unclear, or biased. Our concerns about questionnaires suggest the caution that researchers face in using data from questionnaires in making decisions.

The goal of a questionnaire is to measure respondents' understandings and beliefs in an unbiased manner. An unbiased questionnaire has specifically defined neutral words, rather than using vague words, to avoid provoking and leading the respondent to a particular response. Terms such as "socialist," "out of control," and "family values" are both vague and emotionally provocative. Responses such as "many," "recently," or "sometimes" are vague and create unreliability.

To avoid confusion, the questions should contain positive rather than negative wording. Only one idea should be included in each question, to avoid a situation in which the respondent might agree to one part, but not to the other.

Respondents are often embarrassed to say that they do not know something and may even acquiesce in a response when they do not know the answer. The questions should provide an opportunity for respondents to indicate that they have no knowledge of a situation and that they may disagree with a statement.

The order of questions, a large number of questions, and a limited amount of time may also affect responses. Earlier questions may include information that influences the response to later questions. A large number of questions may contribute to fatigue and carelessness in the respondent. A limited amount of time may also contribute to carelessness as the respondent rushes to complete the questionnaire.

Questionnaires that are submitted and are to be returned by mail have a very low response rate, which may affect the representativeness of the sample of respondents. Those persons who return a questionnaire may have different characteristics than those who do not return a questionnaire. The use of a postage-paid envelope and a short questionnaire increase the rate of return for mailed questionnaires.

The type of response required in a questionnaire affects the obtained results and the interpretation of the results. The following is a list of different types of responses to questionnaire items:

Types of Responses to Questionnaire Items

- The forced-choice method forces the respondent to choose one of the provided options to a question, whether the respondent completely agrees with the option or not.
- The paired comparison method presents a series of two competing options for one of which the respondent must indicate a preference. Different combinations of options are in a sequence so that each option is compared with every other option.
- A Likert scale provides five or seven options from which the respondent selects the degree of agreement or disagreement with the statement. This type of scale typically has a balance, with a neutral point and symmetrical positive and negative degrees of agreement. An assumption is that this type of response uses an interval level of measurement, but some professionals indicate that respondents do not have an equal interval between the different response options. A more modest assumption is that this type of response uses an ordinal level of measurement. This type of scale is the most commonly used in survey research.
- A numerical scale asks the respondent to provide a scaled number, for instance, from a low of 1 to a high of 10, that represents the degree of

agreement with the statement. Respondents assume an equal interval between each number to consider this an interval level of measurement.

- The ranking response method requires the respondent to rank all possible options. Ranking provides an ordinal level of measurement that limits the accurate estimation of degree of response.
- The written statement method requires diligence by the researcher to develop reliable standards for scoring the responses. Whatever the method used to obtain the responses, a sufficient number of categories or options must be provided to capture the full range of possible responses.

An additional important consideration is whether completion of the questionnaire is anonymous. Respondents may provide different answers to questions with their identity revealed than they would if they were anonymous. Would we respond differently if responses were not anonymous?

Sometimes, questionnaires influence the responses for political purposes. This technique is called "push polling," in which the wording or types of responses permitted are constructed to obtain a desired result. Just listen to the popular commentators on cable television and you will hear multiple examples, such as, "Why did the IRS really get direction from the White House to investigate certain groups?" The question presumes that this happened. It is important that we examine each question in the questionnaire for bias.

A recurring theme in collecting data in research is to use multiple types of instruments, such as tests, questionnaires, observation, interviews, physical records, and the like. The researcher will have greater confidence in the results if data gathered from the various instruments provide consistent results and less confidence if different types of research instruments provide conflicting results.

Common Errors

1. Questions are ordered in such a way that earlier questions can affect the responses to later questions.
2. Limited time to complete the questionnaire contributes to careless or inaccurate responses.
3. A tendency to agree with a statement when only presented with two options leads to response bias.
4. A long questionnaire contributes to response bias to later questions because of fatigue.
5. An insufficient number of response categories for pre-coded questions contributes to response bias.
6. Response bias occurs when respondents respond even with a lack of knowledge.

7. Response bias occurs when respondents acquiesce with a socially desirable response.
8. Respondents knowing the sponsor of the questionnaire affects responses.
9. The questionnaire is not completed in person, making it impossible to verify respondents.
10. The low rate of returned mailed questionnaires affects the representativeness of the sample.
11. The identity of respondents is revealed, which may lead to different responses than if identities were kept anonymous.
12. Embarrassing questions on a private issue bias the response.
13. A presuming question affects response.
14. Respondents lack a reference point for responding to a comparison question.
15. Open-ended questions lack clarity of scoring responses.
16. The level of motivation of respondents to complete the questionnaire affects the responses.
17. Hypothetical questions elicit responses that may differ from responses to non-hypothetical questions.
18. Question wording is vague, with words such as "many," "sometimes," or "frequently."
19. Questions contain emotionally provocative wording and labels, such as "socialist," "out of control," or "traditional."
20. Professional jargon or wording is used without reference to meaning.
21. Respondents do not understand complex wording of questions.
22. Questions are negatively or doubly negatively stated.
23. Question wording involves a leading question.
24. Question wording is double-barreled, with two ideas in the same question.
25. Questions involve memory or recall rather than present status.
26. No opportunity is provided to ask for clarification of a question.
27. The forced-choice method of response does not permit determination of the intensity of agreement or disagreement with a statement.
28. The paired comparison method of response does not provide a sufficient number of paired comparisons or a range of values in the comparisons.
29. The Likert method of response does not provide a sufficient number of categories to differentiate agreement or disagreement with a statement.
30. The ranking method of response presents more than a few alternatives, making them difficult to reliably order.
31. The ranking method of response does not indicate the intensity of agreement or disagreement with a statement.

Examples

- Daniel Goldman reported, in *The New York Times*, that respondents differently interpret words even when they have indicated that the words are clear to them. Here is an error of the varying percentage of responses to the meaning of the vague phrase "over the last few years":

    - 14% said no more than two years;
    - 38% said up to seven years; and
    - 41% said 10 or more years.

- *USA Today* reported that a questionnaire that used the word "spending" reinforced the Republican narrative that uses the concept of "tax and spend" as criticism. Respondents agreed that there should be spending cuts, but respondents also said that they wanted no program cuts. The word "spending" is an emotionally provocative word that influences responses and creates an error in interpretation.
- *The Plain Dealer* reported the results of a Pew Research Center study that identified the effects of question wording. Forty percent of people answered yes to this question: "Should laws be passed to eliminate all possibility of special interests giving huge sums of money to candidates?" However, 99% of people said yes to this question: "Do groups have a right to contribute to the candidate they support?" Errors in question wording can affect the responses and lead to errors in interpretation.
- A 2012 Presidential Platform Survey from the Republican Party asked for opinions on "exposing Obama's radical left-wing policies." The three possible options for answers were: Very Important, Somewhat Important, and Not Important. How many different biases and errors are there in that issue statement? (The error is the use of emotionally provocative wording without operational definition of the terms "radical" and "left-wing policies.")

INTERVIEWS

Learning Objectives

1. Understand the different types of interviews, from unstructured to structured questions.
2. Understand how the characteristics of the interviewer affect the responses by the interviewee.
3. Understand how the characteristics of the questions affect the responses by the interviewee.

4. Understand that interviews provide only one way to collect data and that data collected by multiple types of instruments provide greater confidence in the obtained results.

Prerequisite

1. Understand measurement foundations of data collection instruments.

Concepts

Our palms are still sweating from interviews many years ago. How do you feel about interviewing?

Interviews can be a data collection method that permits a dynamic interaction between the interviewer and the interviewee. Have you ever interviewed anyone else? This dynamic interaction, however, creates additional concerns regarding collecting accurate information and avoiding affecting the responses of the interviewee. Unfortunately, many interviewers are poorly trained, which limits the opportunity to obtain accurate information from the interviewee.

A significant issue in collecting data in an interview is to consider the structure of the questions that are used. A structured interview has the same set of questions for each interviewee, while an unstructured interview permits different questions for different interviewees.

The types of responses requested should permit interviewees to provide an accurate response that reflects their full range of knowledge or beliefs. The responses may be limited to a selection of pre-coded options or may permit extensive verbal narrative. The six types of responses described in the previous section in reference to questionnaires are also relevant to the types of responses possible in an interview.

The interviewer must be careful to avoid providing verbal or visual clues that may affect the responses. It is a common occurrence for an interviewee to react to these clues by changing the response to the answer perceived to be preferred by the interviewer. This does not provide accurate information, as the interviewee may simply provide socially desirable answers.

As with questionnaires, the wording and order of the questions, length of the interview, the number of questions, and the type of responses permitted may affect the responses provided by the interviewee. In addition, the characteristics of the interviewer, the decision whether to provide anonymity for the interviewee, and the location of the interview may affect the responses. All of these issues determine the accuracy of the data obtained from an interview.

There is greater confidence in the data if there is agreement with data collected with multiple other types of research instruments. Are we confident in believing what someone says or more confident in observing the behavior?

Common Errors

1. Questions are ordered in such a way that earlier questions can affect the responses to later questions.
2. A tendency to agree with a statement when only presented with two options leads to response bias.
3. Long interviews produce fatigue and bias responses to later questions.
4. An insufficient number of response categories for pre-coded questions contributes to response bias.
5. Agreement with a socially desirable response leads to response bias.
6. Respondents knowing the sponsor of the questionnaire affects responses.
7. The identity of respondents is revealed, which may lead to different responses than if identities were kept anonymous.
8. Embarrassing questions on a private issue bias the response.
9. Presuming questions biases the response.
10. Respondents lack a reference point for responding to a comparison question.
11. Open-ended questions lack clarity as to the length and structure of response.
12. The level of motivation of respondents to complete the questionnaire affects the responses.
13. Hypothetical questions elicit responses that may differ from responses to non-hypothetical questions.
14. Question wording is vague, involving words such as "many" or "sometimes."
15. Emotionally provocative wording and labels are used in the questions, such as "socialist," "out of control," or "traditional."
16. Professional jargon is used without reference to meaning.
17. Respondents do not understand complex wording of questions.
18. Questions are negatively stated.
19. Question wording involves a leading question.
20. Question wording is double-barreled, with two ideas in the same question.
21. Questions involve memory or recall rather than present status.
22. Respondents have no opportunity to ask for clarification of questions.
23. Characteristics of the interviewer (age, gender, ethnicity, appearance) affect interviewee responses.
24. Open-ended questions lack clarity of scoring responses.

25. Interviewer expectations regarding responses to earlier questions affect interpretation of responses to later questions.
26. The degree of probing or asking for clarification of interviewee responses biases or affects interpretation of responses.
27. There are role expectations whereby the interviewer interprets responses in ways expected from the interviewee.
28. An interviewee's level of interest affects the responses.
29. Limited training by the interviewer affects the responses.
30. Pressure to respond leads to a bias of selecting an option easily retrieved from memory.

Examples

- *USA Today* reported that John Gottman's research team interviewed over 2,000 couples and concluded that men need less conflict and women need more friendship. Potential uncontrolled errors or explanations include variations in the degree of training of the interviewers and potential interviewer effect.
- *The New York Times* reported that an annual Gallup poll conducted every year since 1948 to determine which man or woman is most admired has potential biases that affect the responses. The wording and pressure to name a person create a bias and a potential error. Interviewees are most likely to choose the most easily remembered persons.
- *USA Today* reported that the Centers for Disease Control and Prevention and numerous other studies have found low levels of washing hands by medical professionals and the public. A Harris Interactive interview found that 92% of Americans say that they usually or always wash their hands after using the bathroom. Does this high rate reflect a social desirability response bias? Yes: a report by the American Society for Microbiology found that the observed rate of washing hands is approximately 67%.
- *The New York Times* reported that Pew Research conducted a telephone interview survey to determine the degree of support for candidate Barack Obama. A higher degree of support was in the survey than the actual vote because of the "Bradley effect," a response bias wherein a higher percentage of respondents state socially desirable support for a black candidate rather than their actual intended vote. The predicted vote total was in error.
- *The New York Times* reported that a defendant, Sundhe Moses, claimed to have signed a confession only after the detective became physically abusive to him. This is a response bias error of an acquiescent response.

OBSERVATIONS

Learning Objectives

1. Understand that the observer may affect the behavior or performance of the observed persons.
2. Understand the different methods being used to record observations (e.g., frequency of behavior, intensity of behavior, length of observation, type of behavior, antecedent of behavior, consequence of behavior, context in which behavior occurs).
3. Understand that observations provide only one way to collect data.

Prerequisite

1. Understand measurement foundations of data collection instruments.

Concepts

Every day we observe others and others observe us. That is part of life. Have we ever had the feeling of someone watching us? Did that feeling make us uneasy? Did that feeling cause us to change our behavior?

If so, then we understand the dilemma of trying to collect data by observing how others perform or behave while not affecting their behavior. The goal in observation is to collect data regarding how people naturally behave in situations.

Observing the behavior of others is a potentially valuable instrument for collecting data. Listen to comments by television commentators at a football game: "The players have their hands on their knees, they must be getting tired," or "The quarterback came to the line and 'read' the defense." We make conclusions by observing body language such as, "She seems to be angry," or "He's giving me that 'come hither' look."

The observers require training to identify reliably and record the behaviors of concern. Different methods can reduce the effect of the observer. Unobtrusive methods collect data without a person being aware. Techniques of hidden or disguised observations often provide realistic indicators of behavior.

Observation as a data collection tool requires careful planning that involves multiple considerations regarding the types of data to record. Observers can gather useful data on the following characteristics of behavior:

Characteristics of Observed Behavior

- frequency of behavior

- intensity of behavior
- duration of behavior
- type or category of behavior
- antecedents of behavior
- consequences of behavior
- contextual situation

Each of these considerations can affect the data collected and the conclusions made about the observed behavior. The operational definitions will determine which aspects of behavior we observe.

Methods to record observations include recording behavior within a designated time interval or a total of the behaviors over a more extended period. It is important to specify the length of the time interval for observation, as some infrequent behaviors may not occur in a short interval of observation and taking notes.

The recorded observations can vary from anecdotal observations to open-ended ratings or recordings and more formalized rating scales and recordings. More formalized rating and recording methods offer the potential for greater reliability. The training of observers is important to ensure accuracy in recording observations and reliability among observers.

It is important to remember that observations can only be used to collect data about current observable behavior. Some concepts are difficult to observe, such as attitude, intelligence, motivation, or honesty. Different researchers may have different operational definitions of the behaviors that represent these concepts.

Results obtained from observations can provide a comparison to data collected by additional types of data collection methods. Sometimes, people say one thing and do another. Is this something we ourselves have ever done?

Common Errors

1. The observer's presence affects the observed behavior.
2. Behaviors are affected by awareness of potential for observation even if observer is not present.
3. Behaviors observed in artificial situations or contexts do not generalize to a more typical situation.
4. Past behavior is difficult to infer from present behavior.
5. There is difficulty in inferring concepts from observed behavior.
6. Limited training of the observer contributes to inaccurate observations or bias.
7. Recorder observations have any of these biases: generosity, central tendency, severity.

8. Categories to record observations are limited and do not reflect the possible range of observable behaviors.
9. The observer is affected by knowledge of what to expect in observing others.
10. Delay in recording observations contributes to inaccuracy because of deterioration of the memory of the observations.
11. The observer drifts or becomes fatigued from lengthy periods of observation.
12. Observer judgment of behaviors is affected by the general impression or group membership of the observed person.
13. The observed person behaves in a manner that conforms to cues given by the observer.
14. Observer behavior changes because of knowledge of comparison to other observers.
15. There is a primacy effect whereby the initial impression of a person or group observed influences subsequent observations.
16. The performance observed is not an authentic task.
17. The behavior of interest is not directly observable.
18. The performance observed does not have an indication of the time, materials, or other constraints for completing the task.

Examples

- *The Plain Dealer* reported that University of Virginia law professor Brandon Garrett found that faulty eyewitness identification had been the leading cause of errors in convictions of the first 200 convicted defendants exonerated by DNA evidence. Not knowing the error rate of eyewitness testimony creates difficulty in making accurate decisions.
- MSNBC.com reported that the London School of Hygiene and Tropical Medicine found that less than one-third of men and two-thirds of women wash their hands after going to the toilet. Even the idea that there may be an imagined observer creates the error of an observer effect. Twelve percent more men and eleven percent more women washed their hands when they imagined an observer. The error is that many more people claim to wash their hands than actually wash their hands. What measure of hand washing should a researcher use?
- CNN.com reported that the Transportation Security Administration (TSA) employs over 1,000 behavior detection officers to use behavior profiling by observing the behavior of passengers at over 100 airports nationwide. At one airport, Newark Liberty International Airport, there have been over 70 arrests, but most were for carrying fraudulent documents or drug possession. The TSA does not report the number of persons falsely stopped for interrogation based upon incorrect interpretation of observed behavior.

- *Maxim* magazine (noted for publishing figures rather than facts, we are told) published a top 10 list of sex legends in which men indicated the number of women with whom they'd had sex. Here is the number of female partners claimed by these men:

    - Charlie Sheen: 5,000
    - Julio Iglesias: 3,000
    - Jack Nicholson: 2,000

  Do we have to accept their statements made in an interview, or can we directly observe and verify these behaviors? Wow! Now we have your attention. Are some behaviors difficult to observe? Would direct observation change the behavior? Is this a question to investigate? (Relax: We know what you are thinking. The answer is no. One cannot volunteer to be an observer to collect this information. There is a name for that.)
- *The New York Times* reported that Sharron Ivory, a witness to a 1995 murder, says that police coached him into giving false testimony. Ivory claimed that police told him to take a good look at a person with a particular number in a lineup. He assumed that the police knew what they were doing, so he used that as a clue and falsely identified the person. Ivory said, "I figured if it wasn't him, he could beat it at trial."

ARTIFACTS AND RECORDS

Learning Objectives

1. Understand that the operational definitions of whom, what, and how in recorded information may change over time.
2. Understand that interpretation of historical artifacts must acknowledge the discrepancies between current and past human perspectives.
3. Understand that there is limited or no opportunity to check reliability in the collecting and/or recording of historical information.
4. Understand that examination of original documents or artifacts provides potentially more accurate information than examining secondary sources or documents.
5. Understand that there are many types of physical products, such as portfolios, works of art, written compositions, poetry, research papers, and constructed artifacts or products.

Prerequisite

1. Understand measurement foundations of data collection instruments.

CONCEPTS

We all have a history. Some of that history we might want to forget, and other parts we do not accurately recall.

For researchers, the past can provide a rationale or a narrative for our present events. Research involving historical events or data creates difficulties for researchers because of the inability to determine the methods of recording data. Changes in methods of collecting data over time can affect the accuracy of data.

Researchers need to examine original documents and artifacts rather than secondary sources. The original documents may provide a more accurate understanding of the historical event than later interpretations. Reading a document in the original language and finding the earliest known artifact are difficult tasks, but they are important to avoid errors of interpretation.

Here is a thought-provoking question: Since the earliest known biblical citations were many years after the death of Jesus, how do we know that the sayings were not just an interpretation of the teachings of Jesus? A related question is: Since Jesus spoke Aramaic, how sure are we that the translations from Aramaic to Greek to English are accurate? Was anything lost in translation?

Original documents may also create confusion because of changing definitions and terms. Epidemiologists face a challenge in identifying the past rates of diseases and deaths from original death certificates. Consumption, a disease no longer recorded, was a cause of death in early years. The frequency of psychological disorders is difficult to determine. Many early diagnostic categories, such as "neurotic," are abandoned.

We can see the dilemma faced in doing historical research. Ex post facto, or after the fact, explanations provide less control and certainty about the accuracy of the data. There are some ways to try to provide equivalent comparisons, but conclusions based upon historical research are often more tentative than other types of research.

There are many products in our daily lives. It is difficult to imagine a moment in which we are not in contact with a product of some kind, such as the chair on which we sit, the car we drive, and our works of art or literature.

Many types of physical products are available to collect data. Can we use a finished work of art to determine the achievement of an art student? Can we use a portfolio of written essays to determine the progress of a student? Can we use a completed business plan for determining the attainment of a business student? Can we use the success of a surgery in determining the competence of a physician? The answer to all of these questions is yes.

Products can offer a method of collecting data that reveals how a person can apply skills. In addition to evaluating the quality of the finished product, researchers may be interested in the process or sequence of steps used to create the product.

Products can represent an operational definition of a dependent variable. Assessment of products is undertaken with scoring criteria and rubrics that determine the degree of accuracy, completion, quality, length, number of component parts, and many other characteristics. Researchers need to articulate the operational definitions of the criteria and rubrics used in collection of data from products. A continuing controversy is the collection of data regarding works of art. It is erroneous to indicate that works of art are incapable of evaluation. The issue is with the operational definitions of the scoring standards and rubrics used.

Common Errors

1. Secondary rather than original documents or sources of evidence are used.
2. Recording of the event takes place at a long or unknown time after the event.
3. Accuracy of methods used by persons to gather evidence in the original document is unknown or is unreliable.
4. Operational definitions of measures to gather evidence are unknown or imprecise.
5. Potential biases or conflicts of interest in the original source or document are unknown or probable.
6. Operational definitions of measures to gather evidence or prepare artifacts over time are or may be inconsistent.
7. Equipment or materials used to complete the product are of unknown or low reliability.
8. There is variation in evaluation of the product by different persons that may be related to self-selection of the materials or type of product.

Examples

- The Bureau of Economic Analysis made changes in 2013 to the method used to calculate gross national product (GNP). The GNP now includes research and development in arts and entertainment production as business investments. This change raises the calculated GNP and creates an error because of inconsistency for comparison to previous years.
- *The Plain Dealer* reported that Democratic Representative Marcia Fudge of Ohio cited a study by the American Association of University Women that women earn on average 77% of what men earn, based on data from the Bureau of the Census. The Bureau of Labor Statistics reported that

women on average earn 83.6% of what men earn. The difference reflects how each agency measures earnings. The Census Bureau examines annual earnings, while the Bureau of Labor Statistics examines weekly earnings. Errors in interpretation can reflect different operational definitions of earnings. Was a direct comparison made of the earnings between men and women on the same job with the same education and experience?

- Changes have occurred over the years in the *Diagnostic and Statistical Manual of Mental Disorders* (DSM), published by the American Psychiatric Association. The fifth major revision, DSM-V, includes new diagnostic criteria for disorders and some new disorders; it also eliminates some previous disorders. These changes create the possibility of errors in documenting historical or longitudinal trends in disorders.
- *USA Today* reported a grade point average inflation at both the high school and undergraduate levels. Changes and inconsistencies in standards for assigning grades over the years create the potential for errors in documenting historical or longitudinal trends in grade point average.
- CNN.com reported that researchers at the University of Groningen, in the Netherlands, conducted brain scans and found that areas of the brain are activated during an orgasm and different areas of the brain involved in fear are deactivated. When a woman fakes an orgasm, there is no activation of the area of the brain involved in orgasm. However, the area of the brain involved in fear is activated. The physical record as a data collection instrument appears to provide information inconsistent with the interview as a data collection instrument. The error is that different measuring instruments may produce different results. (We can now answer this question: Was it good for him or her?)
- *The New York Times* reported that improved DNA analysis techniques offer greater ability to test physical records and specimens than was previously possible. This improved accuracy permitted the 2013 identification of Albert DeSalvo as the Boston Strangler, linking him to the murder of Mary Sullivan 49 years earlier. The evidence did not change, but the techniques to analyze the physical evidence improved.
- *The Plain Dealer* reported that the Florida Innocence Project found 246 inmates nationwide exonerated by DNA evidence up to the year 2009. The improved physical record instrument has corrected errors in previous convictions. One person, James Bain, spent more than 35 years in prison for a crime that he did not commit.
- Errors in the physical records of immigrants frustrate genealogists. Numerous instances of misspelled names of persons or locations confuse the search for accurate data to research a family history. Are there any errors in one's genealogical records? Is the family name really Rockefeller and was misspelled when our ancestors came here?

REDUCING ERRORS

As illustrated, errors can occur at any stage of the development or application of any type of data collection instrument. Some preventive procedures described in previous chapters continue to be appropriate to reduce errors. Using instruments that have high levels of validity and reliability is necessary to reduce errors. Comparisons to appropriate norm groups or standards are also necessary to prevent errors. The level of measurement precision provided by the data collection instrument must also have sufficient sensitivity to identify appropriate levels of difference in degrees or amounts of a performance or behavior.

A general error prevention strategy is to use multiple types of data collection instruments (both qualitative and quantitative) in a study. Information provided by multiple types of instruments can confirm or refute the information provided by any single type of instrument. If the same result is yielded by multiple types of instruments, then we have greater confidence in the results and a smaller chance of errors in interpretation.

Another general strategy is to examine all obtained evidence for each of the potential errors described in the different chapters of this textbook. A sequential analysis of potential errors offers the most appropriate and comprehensive method to identify errors in obtained data and to prevent errors when planning to gather information. This chapter presented errors with all types of data collection instruments and errors that are specific to each type of instrument.

Many different types of tests are available for use in gathering information, ranging from objective to essay items. Over 40 types of errors frequently occur in these broad areas: inappropriate question wording, inaccurate methods of scoring responses, inclusion of extraneous clues to the answer, inadequate training for test administrators, incorrect answers that have penalties, and many more. Examining the individual items, and the validity and reliability of the tests, is necessary to reduce errors in interpretation.

For questionnaires, more than 30 different types of errors frequently occur in these broad categories: inappropriate question wording, response bias, inappropriate types of response options identifying degrees of agreement or disagreement, low rate of return, and others. Field testing of questionnaires with a small sample of respondents identifies and prevents potential misunderstandings on the part of respondents when conducting research. Every question must be reviewed for potential bias that could affect the results and the interpretation.

The type of response option to questions is an important source of errors in interpretation of the information from both questionnaires and interviews. Results obtained from one type of response option may provide an erroneous indication of degrees of agreement compared to other options for obtaining responses. Comparing the consistency of results among different response options would help to prevent this error in interpretation.

The dynamic relationship between interviewer and interviewee creates the potential for multiple types of errors. Many of the almost 30 potential errors identified during an interview are similar to those reported in the use of questionnaires. Errors with interviews frequently occur in the areas of inappropriate question wording, response bias, mannerisms and behaviors of the interviewer, and other categories. Reduction of errors in information gathered from interviews requires a combination of sufficient training of the interviewer and examination of the question wording and types of response options.

Observation is a valuable data collection instrument because it provides the opportunity to measure actual behavior. Approximately 20 different potential observation errors often occur in these broad areas: observer effect, observer bias, limited categories to record behaviors, and others. Extensive training of the observer, use of unobtrusive or nonreactive observation techniques, and objective and extensive recording methods can all contribute to reduction of errors from observation.

Artifacts and records have a potential advantage in that this information already exists. Errors may be present because of the use of secondary rather than original records, unknown biases in the persons or procedures used to record information, and unknown consistency of procedures used to gather information over time. Using original records and examining multiple records can help to reduce errors by identifying consistent or inconsistent information.

BEST PRACTICES

- Use measures with appropriate levels of validity and reliability.
- Examine individual items in tests, questionnaires, and interviews to identify errors in wording and question construction described in this chapter (e.g., bias, leading question, presuming question, question involving recall by respondent, emotionally provocative question, double-barreled question, negative phrasing of question, hypothetical question, vague wording).
- Avoid penalizing respondents for incorrect or omitted answers in tests.
- Use objective items, forced choice, and various rating or ranking methods for measuring basic opinions and knowledge in tests, questionnaires, and interviews.
- Use essay items and more open-ended response formats for measuring higher-order or more complex opinions and knowledge in tests, questionnaires, and interviews.

- Use anonymous responses in questionnaires to minimize social desirability.
- Use caution when interpreting results from mailed questionnaires with less than 50% return rate.
- Use unobtrusive observation techniques to minimize observer effect.
- Look for clear indications of what is observed and recorded, with preference for behaviors that have quantifiable characteristics (e.g., frequency, intensity, duration, type or category, antecedents, consequences, context).
- Obtain and examine original artifacts rather than secondary or derived artifacts.
- Provide training to ensure that test administrators, observers, and interviewers have high levels of reliability in administering and scoring tests, observations, interviews, and questionnaires.
- Field-test all tests, observation systems, interview systems, and questionnaires to detect and correct any errors.
- Use multiple types of data collection instruments to provide confirmation of results.

INTERRELATION TO OTHER TOPICS

We have examined different types of evidence collection instruments used in research in the past six chapters. Errors can occur with each type of instrument. A useful strategy to minimize errors is to use multiple types of evidence collection instruments. We will have greater confidence in the accuracy of the evidence if the different types of evidence collection instruments confirm the same result.

The next chapter describes how we obtain the people with whom we conduct the research. We cannot always conduct research on everyone we want to investigate, but we can get a sample of that population. The goal is that the sample accurately represents the larger group of people we wish to study.

Chapter Seven

Sampling Methods

"The Chosen Ones"

By a small sample we may judge of the whole piece.

—Cervantes

We know in part, and we prophesize in part.

—The Holy Bible

I am a part of all that I have met.

—Lord Tennyson

O World, thou choosest not the better part!

—George Santayana

SIGNIFICANCE

We often want to conduct a research study in the hope that the results can be applied or generalized to a larger group of persons or situations. The population is the entire group or universe about which we want to generalize the research results. Populations can differ by characteristic and the number of persons or situations that have that characteristic.

We are rarely fortunate enough to have the opportunity to gather evidence from the entire population about which we want to make a decision. Limitations of time, resources, and availability prevent gathering data from larger groups of persons or situations. Therefore, we must carefully select some who will accurately represent the larger group. A sample is a part of that larger group.

Errors in obtaining a representative sample distort the accuracy in generalizing or applying the results, causing erroneous conclusions about what will happen in the population of interest.

Different methods of obtaining a sample are included in this chapter, and each method varies regarding its potential ability to obtain a representative sample. The method used to obtain the sample in a research study can affect the accuracy of the evidence.

LEARNING OBJECTIVES

1. Understand the concept of a random sample.
2. Understand the concept of a stratified random sample.
3. Understand the concept of systematic random sample.
4. Understand the concept of a cluster sample.
5. Understand the advantages and disadvantage of random, stratified random, systematic random, and cluster samples.
6. Understand the concept of an operational definition for the strata in a stratified random sample.
7. Understand how to obtain a random, stratified random, systematic random, and cluster sample.
8. Understand the limitations or inadequacies of convenience, multistage, and quota sampling.
9. Understand the concept of representativeness in obtaining a sample.
10. Understand the limitations of representativeness from a small sample, using volunteers, or a high rate of refusal by participants.

PREREQUISITES

1. Develop research questions that identify the people on whom the research information will be collected.
2. Apply the concept of operational definitions.

CONCEPTS

We make countless decisions during our lifetime. In many situations, we have to make a decision based upon very limited information, such as a first impression, the comments from one person, or a brief observation. Just listen to the everyday phrases regarding sampling: "Sample-size package"; "A trial issue"; "Take a test drive"; "Three-minute speed date"; "What a motley crew"; "Take your pick of the litter"; "Any volunteers?"; "Love at first sight."

How many times have we found that our opinion of someone based upon this limited information changed after we had more information? What we needed was a better sample that more accurately represents the entirety of the person or situation. The sample must accurately represent the larger group or the population of interest. If the sample is not representative, then decisions may be inaccurate. Examination of a research report includes the identification of any errors in selecting a representative sample.

Selecting a representative sample requires consideration of issues that can affect the results. The first issue is clearly indicating the characteristics of the population to which the results generalize. Demographic characteristics can include, but are not limited to, socioeconomic status, residence, age, political affiliation, occupation, and gender. The goal is that the sample represents the same proportion of that characteristic as the population. An unrepresentative sample creates errors in generalizing the results to the population of interest. Specification of the characteristics of the sample is required.

A second issue is the size of the sample. In general, a larger sample can have a greater chance of representativeness, because more of the population is included. Practical limitations of time, resources, and availability can restrict the sample size. A small sample may be insufficient to represent the population and may contribute to erroneous conclusions about the population.

A third issue is the use of volunteers for the sample, since persons who volunteer often respond differently than those who do not—and this difference creates an unrepresentative sample. Volunteers create a biased sample and errors in generalizing to a population of non-volunteers.

A fourth issue is rate of refusal in which a high proportion of those selected refuse to be part of the sample. Those who refuse to participate may have unknown differences that set them apart from those who agree to participate. There is a potential for bias or error in generalizing the results of a research study that had a high proportion of persons who refused to participate.

Four methods of obtaining a representative sample provide some degree of appropriate representativeness.

Representative Sampling Methods

1. Random sampling: Everyone in the larger group about whom we want to make a decision has an equal and random chance of selection. An advantage of this method is the relative ease in obtaining a sample; a disadvantage is that the irregularities of randomness may contribute to a lack of representativeness unless the sample is large.
2. Stratified random sampling: Each person in the population of interest is located and identified on relevant characteristics or strata. These

strata may include but are not limited to age, intelligence, and occupation. Random selection of individuals with those characteristics then creates a sample that includes the same proportion of people with those strata as the larger group about whom we want to make a decision. This method has the highest probability of obtaining a representative sample with the smallest sample compared to other methods. The disadvantages are the time and difficulty involved in identifying the strata of each person.

3. Systematic random sampling: This method requires creating a list of all members in the population of interest. A starting point on that list is randomly selected for a systematic selection, such as every 10th member of the population. The advantage is the relative ease in selecting the sample. The disadvantage is that irregularities in the listing may create an unrepresentative sample.
4. Cluster sampling: The three methods above select all individuals to be in the sample. Cluster sampling selects existing groups, rather than individuals, to be in the sample. Each randomly selected group has an equal chance of being in the sample. Cluster sampling has the advantage of ease of selecting existing groups. The disadvantage is the irregularities resulting from differences among existing groups. The implication is that the sample size must be larger than in the other three methods to have the same degree of representativeness.

Unrepresentative Sampling Methods

Other sampling methods used in research studies have serious limitations that can lead to an unrepresentative sample and contribute to errors in generalizing the results. The following three methods should generally be avoided:

1. Convenience sampling selects whomever is available. The advantage is that it is easier and less costly to select those individuals. The disadvantage is that there is no assurance of the degree of representativeness of the sample.
2. Multistage participant sampling selects individuals recommended by those initially selected. The assumption is that those initially selected will know of others with the same characteristics. The advantage of this method is the ease of selection. The disadvantage is the unknown probability of the representativeness of the sample.
3. Quota sampling is a nonrandom method whereby individuals are selected if they have a specific characteristic or strata. The advantage is that the sample can have the same proportion of persons with a specific characteristic. The disadvantage is that the sample may not be representative because of bias or error of nonrandom selection of persons.

Identifying a sample of people with very particular characteristics is an important new technique called microtargeting and is used in political advertising. The Michigan Republican Party has sorted each of the millions of voters in the state into 42 categories, based upon easily available data: age, income, gender, voting behavior, magazine subscriptions, vacation destinations, automobiles, and many others. Persons most likely to be receptive to the message receive the political ad at less cost than sending it to all voters.

Microtargeting is customer profile modeling in business or commercial applications. A more precise message or advertisement can be constructed and then sent to people with particular characteristics.

The growth in the number of cable television news and talk radio stations reflects the targeting and stratification of populations with particular political beliefs. These stations can then set higher advertising rates because of representativeness of their audience as a sample of a population of interest. The danger of this increased microtargeting is that these television and radio shows have less representativeness in generalizing to a broader population.

Some companies make money just by selling customer lists for 3 to 20 cents or more a name. Subscribe to an upscale magazine just for fun to learn who sells names. Place your middle initial or a title, such as Mr. or Ms., that you do not often use on that subscription. If you start to receive advertising with that initial or title, then congratulations. You are now part of a stratum in the population. This could lead to the use of a more contemporary pickup line, like "What's your stratum?" rather than "What's your sign?" (We probably need to do research to determine if this new pickup line is more effective!)

COMMON ERRORS

1. The sample is not proportionately representative of the population of interest.
2. Small sample size creates an unrepresentative sample.
3. The proportion of persons refusing to participate is large or unknown.
4. The study uses volunteers, who may not represent the characteristics of others.
5. The rate of withdrawal from participation is high or unreported.
6. Participants with certain characteristics have a differential rate of withdrawal.
7. Participants' level of motivation to participate is unknown or can affect responses.
8. Inaccurate procedures are used for obtaining a random, stratified random, systematic random, or cluster sample.

9. Inappropriate operational definitions are used for strata of participants in a stratified random sample.
10. There is a lack of equivalence in the groups selected using cluster sampling.
11. A convenience, multistage, or quota sampling rather than a representative sampling is used.
12. The proportion of participants with specific characteristics is not indicated.

EXAMPLES

- *The New York Times* cited a study by Dr. Andrew Wakefield that used a sample of only 12 children with fabricated data to declare that vaccinations contribute to risk of autism. No other researcher has replicated the results. A consequence is that many parents have refused to have their children vaccinated, with a resulting increase in the number of children with measles. (A positive result of this erroneous research is a real shot in the arm for identifying errors in research. Will television talk show hosts report the increase in childhood disease?)
- *The Plain Dealer* reported that Dr. Joseph T. Lurito of Indiana University conducted a brain scan study of 10 men and 10 women and found that men listen mostly with the left side of the brain and women use both sides of the brain. No details were included about the characteristics of the sample and the selection of the sample. The error lies in selecting a small and unrepresentative sample of men and women. (We do not know about others, but we are getting tired of all this research about men. We have half a mind to complain about it.)
- *The Plain Dealer* reported that some Greater Cleveland suburban school districts refused to allow high school students to participate in a survey about smoking, drugs, alcohol, sexual activity, and other risky behaviors. This decision creates an unrepresentative sample because of the high differential rate of school districts refusing to participate.
- *USA Today* reported that Glenn Beck erroneously estimated the size of a crowd at a Washington, DC, rally a few years ago. Professionals used aerial photography with high resolution and estimated the crowd at only a fraction of the size estimated by Beck. The high estimate created an error of representativeness of the crowd.
- A classic example of an unrepresentative sample was the prediction that Franklin D. Roosevelt would lose the presidential election in 1936. A random sample of voters was selected from the phone book. The error was that only the wealthy had phones during the Great Depression, and the wealthy were not representative of the entire voting population.

- Motoko Rich reported, in *The New York Times*, that public school students in 47 states achieved higher average scores on the Trends in International Mathematics and Science Study (TIMSS) science test than the international average. The tests claimed administration to a random selection of representative students. The conclusion is erroneous because the international student sample was unrepresentative, including many developing countries and excluding many developed countries.
- Dana Milbank reported, in *The Plain Dealer*, that television personality Sean Hannity had a sample of six "average Americans" who claimed adverse consequences of Obamacare. One of the sample indicated that the law prevents him from hiring full-time employees because he would have to provide health insurance. His business is so small as to not be included in the requirement. Is this misunderstanding representative of the population? Is this a biased sample, comprising only those with a particular view? This is also an error of a small unrepresentative sample.

REDUCING ERRORS

Using appropriate methods of obtaining a sample can reduce the frequency of erroneously generalizing results to a population of interest. The following procedures can prevent or reduce sampling errors:

- First, specifically identify the demographic or situational characteristics of the population of interest.
- Second, identify the proportion of the population of interest with those demographic or situational characteristics.
- Third, use an available sampling method that minimizes bias and provides the highest probability of obtaining a representative sample. The constraints of the situation, however, may limit the sampling method that is used.
- Fourth, avoid the use of volunteers if at all possible, as they may behave differently than the population of interest.
- Fifth, obtain a sample that is large enough to be representative for each characteristic or strata.
- Sixth, indicate the rate of refusal to participate. A high rate indicates potential concern about representativeness of the sample.

Stratified random sampling offers the highest probability of representativeness and preventing errors in generalizing to a population. Professional groups that conduct polls prefer stratified random sampling because of greater assurance of representativeness. Extensive data banks contain numerous personally identifiable characteristics or strata about each of us. This information can assist in identifying an accurate stratified random sample, but at what cost to personal privacy? How many strata do we represent? Which strata identify us?

Random, systematic random, and cluster sampling methods also offer some potential for reducing errors. However, these methods require the use of a larger sample size to offer the same probability as a stratified random sample of obtaining an accurate representation of the population of interest. The greater probability of erroneous representation of the population limits the interpretation of results for simple random and cluster samples..

Other sampling methods presented in this chapter (i.e., convenience, multistage, quota) should generally be avoided. These methods provide even weaker assurances that the obtained sample accurately represents the population. The use of appropriate sampling procedures can reduce, though not prevent, erroneous conclusions reached because of unrepresentative samples.

BEST PRACTICES

- Clearly define demographic characteristics of the population of interest.
- Use stratified random sampling whenever possible.
- Use random, systematic random, or cluster sampling when unable to use stratified random sampling.
- Avoid unrepresentative sampling methods.
- The size of the sample should be sufficient for the population and the subgroups to which results are generalized (generally 15 or more for each subgroup).
- Avoid use of volunteers.
- Use caution when interpreting results in studies with high rates of nonresponse or withdrawal (more than 50% nonresponse or 20% withdrawal) and use matching or other techniques to equate groups.
- Generalize results and recommendations only to a population represented by the sample.

INTERRELATION TO OTHER TOPICS

This chapter discussed the importance of reducing errors by obtaining a sample that accurately represents the population of interest. Common errors and examples were included to illustrate the importance of obtaining an accurate sample.

The real world can be complicated, with many simultaneous events and influences or variables. Even in an accurate sample, errors may still happen because of uncontrolled variables within the specific research study. The next chapter presents many other potential explanations for the obtained research results. This reinforces the idea that research deals with probabilities rather than certainties.

Chapter Eight

External and Internal Validity Threats to Evidence

"Bad Hair Day"

We made our decisions in the light of those values. Yet we were wrong, terribly wrong. We owe it to future generations to explain why.

—Robert McNamara

"Murphy's Law": If something can go wrong, it will.

—Anonymous

To generalize is to be an idiot.

—William Blake

To some extent, if you've seen one city slum you've seen them all.

—Spiro Agnew

It is tempting, if the only tool that you have is a hammer, to treat everything as if it were a nail.

—Abraham Maslow

SIGNIFICANCE

A goal of research is to gather information that demonstrates the effects of the research manipulations done in the study or the accuracy in predicting performance. Errors involved in any of the procedures described in this textbook can limit the ability to reach this goal and can contribute to erroneous decision making.

This chapter provides a description of uncontrolled variables that may influence the effect or the accuracy of the prediction. The possibility of making an erroneous conclusion or decision is greater whenever it is more likely that these uncontrolled variables influence the effect or prediction. As repeatedly illustrated in this textbook, errors can have significant adverse consequences.

Chapter 4 described independent, extraneous, and dependent variables. This chapter describes multiple types of extraneous or uncontrolled explanations. The alarming realization is that each of these potential explanations is present for whatever research information is gathered. Controlling for these potential explanations is an important concern and a significant task for researchers.

Approximately 50 different alternative explanations are included in this chapter. Uncontrolled alternative explanations are errors of or threats to the validity or accuracy of conclusions. Space constraints allow presentation of only selected examples of these explanations.

Uncontrolled alternative explanations that distort the accuracy of the results of the specific research study are threats to or errors of internal validity. Those explanations that distort the accuracy of generalizing results beyond the study to other persons, situations, or measures are threats to or errors of external validity.

Unfortunately, many research studies provide limited control of these alternative explanations. Identifying these alternative explanations can assist in reducing errors in interpreting research information. These explanations exist in real life and have real-life implications for making decisions.

LEARNING OBJECTIVES

1. Understand that internal validity is confidence in the accuracy of the results of a research study.
2. Understand the many other potential explanations or threats to the internal validity of a research study.
3. Understand that interpreting the results of research requires an examination of other potential explanations.
4. Understand that external validity means extension of the results of a study to different persons and/or contexts.
5. Understand that other explanations or threats limit the generalizability of results from a study to other persons and/or contexts.

PREREQUISITE

1. Understand the meaning of these types of variables:

   a. Extraneous
   b. Mediating
   c. Moderating
   d. Confounding
   e. Multi-collinear

CONCEPTS

We gather information in our daily lives in order to make decisions about what causes something to happen or to predict what will happen. We want to be certain that the results that we observe or predict are not due to other potential causes or variables. If other potential causes or explanations are plausible, then we will have less confidence and accuracy in making a decision.

Researchers use the term "internal validity" to refer to the extent to which other extraneous or uncontrolled variables influence the change in or the prediction of the dependent variable. The greater the control of these other extraneous variables, the higher the internal validity and the confidence in the results.

As described in chapter 4, many potential variables affect the accuracy of a study. These variables mediate, moderate, confound, or vary together with other variables. The following are some of the important variables that may affect the internal validity of a study:

- demographic characteristics of participants (e.g., age, gender, ethnicity, education, political belief, socioeconomic status)
- attitude of participants
- participants who are volunteers
- excessive and/or differential withdrawal of participants
- historical events
- location of the research study
- inaccurate implementation of research treatments
- pretesting
- unrepresentative sample
- inadequate time to demonstrate effects
- use of extreme scores

All of us have experienced the frustration that occurs when something works for someone else or in a different situation, but does not work for us. "External validity" means that the results obtained by other persons—or in other situations—should also work for us. However, there may be other explanations for why the results do not generalize. These other explanations are threats to external validity.

Research results have limited use without generalization to other contexts that are external to a specific study. We do not constantly have to reinvent the wheel if knowledge can be generalized to other situations.

Think for a moment about how a physician makes a diagnosis. Previous research showed a pattern of symptoms in many other persons and medications that effectively treated the symptoms. If the past research has external validity, then that pattern and prescription should generalize to you. What if the past research does not generalize to our situation? A researcher would say that there is no external validity to the result, and that there may be errors or threats to the external validity. We would say OMG and hope that we have good enough insurance to obtain a second opinion.

The context or the location in which the research is conducted, how the persons were selected to be in the study, and inappropriate generalization of the results to other groups are among the most common errors that adversely affect external validity. Careful consideration of how the investigation is conducted and the control of other extraneous variables can increase the external validity of a single investigation.

COMMON ERRORS

1. Subject characteristics affect the results of the study.
2. Extremely low or high scores cluster closer to the average on retesting regardless of treatments in the research study.
3. Mortality leads to a loss or withdrawal of persons in the study, creating a smaller and/or unrepresentative sample.
4. Instrument decay causes deterioration in accuracy of equipment used or fatigue and drift in accuracy of the provider of data.
5. Data collector characteristics such as gender or age affect responses.
6. The data collector unknowingly alters data to make a particular outcome appear more likely.
7. Culturally biased questions or instruments affect results.
8. Data collection instruments lack adequate validity, reliability, and/or differentiation of degrees of response.
9. The type of data collection instrument affects the responses.
10. The format of the instrument (selects a response, constructs a response) affects the response.

11. The response is affected by whether the identity of the respondent is or is not anonymous.
12. Responses are different when the instrument is administered individually or in groups.
13. The location of data collection affects responses.
14. Pretesting provides practice that improves later responses to the data collection instrument.
15. Responses reflect maturation in age, physical development, or experience regardless of treatments in the study.
16. Events during the study affect the responses of all persons or differentially affect some persons.
17. Subject attitudes change in that some persons perform better because of novelty of treatment or special attention (the Hawthorne effect).
18. Subject attitudes change in that some persons in a control group perceive a lack of special attention and perform more poorly.
19. Subject attitudes change in that some persons in a control group know that they are in an experiment and try to outperform other groups (the John Henry effect).
20. Volunteer subjects are not representative of attitudes of others and respond positively to any treatment.
21. Subjects disagree with treatment to be administered and respond poorly to the treatment.
22. Some persons do not implement the treatment as intended (lack of treatment fidelity).
23. Persons in different groups receive different treatment replications (treatment exposure).
24. Persons in one group are exposed to persons in another group who receive a different type or amount of treatment.
25. Persons are given more than one treatment, resulting in unintentional interaction (multiple treatment interference).
26. The implementer of treatment becomes more proficient with each opportunity to practice, regardless of the specific treatment or intervention.
27. Persons expect to improve regardless of treatment because of a placebo.
28. The type of placebo used as a treatment affects response (e.g., a more expensive placebo is better than a less expensive one).
29. A study of brief duration does not permit demonstration of latent effects.
30. An excessively long study permits the influence of additional factors to minimize the treatment effects.
31. Results are affected by inappropriate start and end dates of study.

32. Treatments are sequenced such that earlier treatments affect how persons respond to later treatments.
33. Non-random selection causes interaction of selection and treatment.
34. Pretest-treatment interaction sensitizes persons to treatment compared to those who are not pretested.
35. Results are inappropriately generalized to locations other than the location of the study.
36. Results are inappropriately generalized to other contexts with different operational definitions.
37. Results are inappropriately generalized to other types of persons (subject extrapolation).
38. The effects of independent variables are inappropriately extrapolated and generalized beyond the amounts or types investigated.
39. The effects of particular independent variables are inappropriately generalized to intermediate amounts or types of the variables investigated.
40. The effects of one independent variable are inappropriately generalized when persons were exposed to multiple independent variables (multiple independent variable interference).
41. The effects indicated in the dependent variables are inappropriately generalized to other types of dependent variables or to other measures of the dependent variables (dependent variable extrapolation).
42. Measurement of one dependent variable affects the measurement of other dependent variables.
43. Inappropriate generalization occurs because performance on prior dependent variables affects performance on subsequent dependent variables (dependent variable contamination).
44. Inappropriate generalization occurs when repeated measurement on the same dependent variable during the research study is not representative of the natural setting (multiple measurement artificiality).
45. Results are inappropriately generalized from an artificial or reactive experimental setting to a natural setting.
46. Results collected at one time are inappropriately generalized to other times.
47. Results from a study in which a control group did not receive an alternate or compensatory treatment are inappropriately generalized to other situations in which a control group did receive an alternate or compensatory treatment.
48. Results of a study are inappropriately generalized to other situations with a different initial baseline.
49. Results from a single research method are inappropriately generalized to other research methods.

EXAMPLES

- *The Plain Dealer* reported that George Gallup conducted a study comparing women whose partners wear condoms with those whose partners do not wear condoms. The results indicated that women directly exposed to semen during sex are happier than women who are not directly exposed. Is it an error to conclude that semen makes women happier? Is happiness because of the hormones in semen? Is another explanation the subject characteristics of women with partners who do not practice safe sex? (The interpretation of happiness could change after collection of data about the additional dependent variable of contracting a sexually transmitted disease from unprotected sex.)
- *The New York Times* reported that the Bureau of the Census changed the definition of "nuclear family" to exclude when another adult—besides the biological parents—lives in the home. Other historical inconsistencies over time did not count the family as a traditional nuclear family if the mother worked outside the home for even a few hours a week. These historical inconsistencies create potential errors in longitudinally determining the number of nuclear families over a period of time.
- *The Plain Dealer* reported that in 2005, the Environmental Protection Agency was required to adopt a shorter reporting form for companies to disclose spills and releases of toxic substances. The new standard exempted companies if there were less than 5,000 pounds of a specific chemical. The previous standard was 500 pounds. This historical event creates a threat to the validity of examining longitudinal changes over time. (Just want to make sure we clear the air on this issue.)
- The *British Medical Journal* reported that men who had sex at least twice a week had a death rate half that of men who had sex less than once a month. Besides health, what other extraneous demographic subject characteristics exist? (No data were included for the longevity of men who had frequent sex with another man's significant other.)
- *The Plain Dealer* reported that Earl Smith and Li Fang Hung of the University of Houston and M. Crawford of the University of Texas published a study in *Nature Medicine* claiming wearing glasses to correct a vision problem might make a child's vision worse. That study used 11 infant rhesus monkeys who wore headgear fitted with different lenses. This error is in the validity of generalizing results to humans. Think of how many parents erroneously refuse to have their child wear corrective lenses. The National Institutes of Health has now placed a moratorium on new studies using chimpanzees. (We think the real reason for that is they were concerned with all of the monkeying around in the research laboratories.)
- *USA Today* reported that former Florida Governor Jeb Bush asserted a significant increase in the proportion of fourth-grade students in Florida

that had passed the proficiency test because of his retention policy and a focus on basics. Maturation of retained students—who were a year older upon retesting and had the additional year of practice on the same test topics—prevents generalization.

- According to *The New York Times*, The Heritage Foundation stated that those living below the poverty line in America are not poor because they have microwaves and refrigerators. Does this operational definition of poverty indicate an error in generalizing the concept of poverty to other contexts and to other operational definitions of poverty? What is the appropriate norm group for comparison?
- *The New York Times* reported that the National Resource Center on Child Sexual Abuse found significant variability in data regarding the frequency of sexual abuse among different studies. No uniform method of collecting the data exists, and different studies have different operational definitions of sexual abuse. Increases in reports of abuse in the past 30 years may reflect greater recognition of abuse rather than an actual increase in abuse, as indicated by data showing that 60% of reports are unsubstantiated compared to 35% of reports 30 years earlier. Different operational definitions and increased recognition may create errors in external validation found in other contexts and times.
- Michael Gerson reported, in *The Plain Dealer*, that short-term trends in the environment exaggerate results based upon the choice of a statistical starting point. A graph beginning in 2000 would show a different slope of temperature change than a graph beginning in 1998. Different starting points may contribute to errors in interpretation of longitudinal data.

REDUCING ERRORS

The vast number of uncontrolled alternative explanations presented in this chapter creates a profound dilemma for researchers and for consumers of research information. Almost daily, a new research study contradicts the results of a previous research study. A prime reason for these contradictory results is the lack of control due to the influence of alternative explanations, as each research situation may present a different mix of uncontrolled alternative explanations.

The frustration of consumers with contradictory research information is obvious. As a result, many will assume that research can show whatever result the researcher wants.

Identifying these alternative explanations is important to understand the limitations of the information and to reduce errors in interpreting the information. It may be impossible to prevent all errors in a research study, but it is possible to acknowledge limitations and present alternative explanations.

Reducing errors may require additional information. Research designs or procedures that provide more control of the alternative explanations reduce errors. Research then becomes a process by which each research study builds upon the results of previous research and examines additional potential alternative explanations. The next chapter includes techniques to reduce errors from uncontrolled alternative explanations.

BEST PRACTICES

- Identify potential threats to external and internal validity described in this chapter.
- Use any of the techniques described in chapter 9 to minimize threats to validity (e.g., randomization, matching, subject characteristics as a constant, subgroups, repeated treatment, covariance analysis, single-blind design, double-blind design, subject characteristic as a variable, multiple experimenters, more valid and reliable instruments, multiple instruments, placebo group, control group continuing current treatment, treatment fidelity with continuous monitoring of treatment implementation).
- Select a research design that minimizes threats to validity.

INTERRELATION TO OTHER TOPICS

This chapter indicated the limited usefulness of research if we see other probable explanations for the results or if we cannot generalize or apply research results to other persons or contexts. A research study must have strong internal and external validity to reduce the probability of errors.

The possibility of other explanations reinforces the concept that research only reaches tentative conclusions. Researchers try to minimize these other explanations with the use of appropriate research procedures, but we must humbly acknowledge other possible explanations. Research is a dynamic process of continual investigation of other or more plausible explanations for the results of a study.

The next chapter examines issues involving plans or designs for conducting research comparing individuals or groups. These designs offer varying degrees of control of the alternative explanations that can affect both internal and external validity.

Chapter Nine

Research Design

"Feng Shui of Research"

You know my methods, Watson.

—Sir Arthur Conan Doyle

What I discovered was the scientific method by which the unconscious can be studied.

—Sigmund Freud

The country demands bold, persistent experimentation. It is common sense to take a method and try it. If it fails, admit it frankly and try another.

—Franklin Delano Roosevelt

Make no little plans; they have no magic to stir men's blood.

—Daniel Burnham

SIGNIFICANCE

Research designs encompass the overall structure, variables, procedures, and methods used to collect and analyze information comparing individuals or groups. These designs vary in degree of control of alternative explanations and appropriateness for different conditions and purposes of investigation. Some designs permit direct manipulation and control. Other designs do not permit direct manipulation and control as the event or behavior has already occurred. In this situation, the research can only describe, discover relationships, or attempt to infer a cause after the fact.

The specific type of research design that is used provides both the direction for gathering information and limitations in interpreting the information. Errors can occur if designs are used that provide limited or no control of extraneous or alternative explanations. Further errors can occur when conclusions are made beyond the limitations of the research design that is used.

Chapter 3 described different types of research questions. The research design that is used should be appropriate for the question and the constraints in the research situation. This chapter includes common errors in design and representative examples of errors.

LEARNING OBJECTIVES

1. Understand the different methods by which to design and conduct research.
2. Understand that different methods are appropriate for different types of research questions.
3. Understand the multiple techniques used to control or minimize the effects of threats to internal and external validity.
4. Understand that interpretation of research information considers limitations for any potential errors or threats to the validity of the research study.
5. Understand that interpretation of research information includes suggestions for direction of future research.

PREREQUISITES

1. Understand the concepts of independent and dependent variables.
2. Understand the concepts of internal and external validity.
3. Understand the threats to internal and external validity.

CONCEPTS

We all have dreams and goals, which can be as varied as getting an education, saving for a home or retirement, or having a satisfying relationship with a significant other. We (a) make a plan to reach those goals, (b) decide what information to collect, (c) identify what actions to take, and (d) determine whether we have achieved our goal. "Research design" is the term used by researchers to describe the similar overall plan to achieve the goal of the research investigation.

The quality of research depends upon appropriate practices in each of the component skills of research described in this book. Weaknesses or errors made in any of the components can adversely affect the accuracy of the results and decisions made. Research is only as good as the weakest link in the process.

Many different types of research designs provide evidence that supports or refutes a hypothesis. Only a brief discussion of all of these designs is possible in this introductory textbook. However, a foundational knowledge of the types of questions answered and the limitations of each type of design is necessary to understand potential errors.

Types of Research Designs

- *Descriptive designs* attempt to provide a summary of the frequencies or variability of scores on a variable. The purpose of the design is to provide the status of the behavior of a group. There is no attempt to indicate a relationship or to infer causality. An error is attempting to generalize the results to other groups or proposing a cause-effect relationship.
- *Correlation and prediction designs* seek to identify the strength and the direction of a relationship between two or more variables. These variables can be from the same individuals at different times or scores from different data collection instruments. The variables may also be measures or scores from different individuals. Without manipulation and control, the direction and strength of the relationship are calculated. A purpose for this design is to predict the scores or categories for one or more variables based on the scores or categories obtained for other variables.

  Limitations of the correlation design may contribute to errors. The first is the assumption of a constant or linear relationship across the entire range of scores for the variables. A second limitation is the assumption that the entire range of possible scores or categories of variables is used. A truncated range of scores or categories requires a modified method of statistical analysis.

  Perhaps the most significant error in correlation design is one of interpretation. No matter how strong and consistent the obtained correlation, correlation does not indicate causation. Other uncontrolled variables may contribute to this relationship.
- *Comparative and causal-comparative designs* seek to determine differences between groups on one or more dependent variables under conditions in which the researcher has no direct control of an independent variable or treatment. Comparative designs do not attempt to infer a causal relationship.

  Causal-comparative designs do not have direct control of a potential cause. Comparisons among groups attempt to identify any differences in response to an intervention while it is happening. The lack of direct control of the assumed cause contributes to errors in interpreting causality.

Matching and using subject characteristics as a constant are techniques for minimizing errors.

- *Ex post facto designs* also attempt to infer causality by determining if there are significant differences among groups when there is no direct control of the assumed cause. The ex post facto design examines a potential cause or intervention that has already happened, while the causal-comparative design may examine the potential cause or intervention as it is happening. The lack of control in the ex post facto design contributes to the same potential errors in interpreting causality. Matching and statistical adjustment can minimize some other explanations.
- *Quasi-experimental designs* do not have assurance that comparison groups are equal at the time of manipulation. They examine whether groups that will receive a different manipulation of a characteristic will perform differently on another characteristic. Quasi-experimental designs address questions such as these: Will sales be higher when a group receives an advertisement with a celebrity endorsement than when a group receives an advertisement without a celebrity endorsement? Will students have higher test scores in an online class than students in a traditional class?
- *Experimental research designs* offer the possibility of direct control of potential causes and greater control of other potential explanations or threats to external and internal validity. Assignment of individuals to form equivalent comparison groups occurs before administration of an assumed cause or intervention. Manipulation of a potential cause or causes determines the effects. Specific experimental research designs vary in their degree of control of the threats to external and internal validity and the confidence in the accuracy of the obtained information. Experimental designs may incorporate multiple variations in manipulation that have implications for the degree of control and for the comprehensiveness of the obtained information.
- *Single-subject designs* investigate only a single person. The purpose of a single-subject design is to determine a person's response to the administration and withdrawal of a treatment. This design is often used in therapy with persons with special needs. Other designs investigate groups of persons and can be descriptive, correlational, causal-comparative, ex post facto, experimental, ethnographic, mixed-method, and many other types.

A complete listing of these variations is beyond the scope of this introductory textbook. However, some variations are common, and students need to understand the reasons behind and limitations of these variations.

Research Design Considerations

- The first consideration is the number of persons receiving each treatment or combination of treatments. A greater number of persons receiving a treatment generally provides greater confidence in the results.
- Second is the type of variable or variables manipulated. These variables can be categories or types, such as gender. The variables can also represent different amounts of, for example, time or money. Variables manipulated by amounts or degrees allow for finer discernment of the effects of the potential cause or causes.
- Third is the number of potential causes manipulated in one research study. Manipulation of two or more potential causes in one study is factorial and has the advantage of identifying the separate effects of each potential cause and the combined effect or interaction. An interaction effect can be determined only when all combinations of different categories or degrees of the manipulated variables or potential causes are included.
- Fourth is whether pretesting occurs before administration of the potential cause. Pretesting may sensitize the individuals to the potential cause or data collection measures and is a threat to the validity of the study. The use of unobtrusive measures or an additional comparison with those individuals who were not pretested may minimize this concern.
- Fifth is whether each individual receives one or all combinations of the potential causes or treatments. Each person receives all combinations of the treatments in a repeated measures design. The assumption with this design is that there is no initial difference among groups of individuals, since everyone receives all combinations of the treatments. The limitation is that the order of administration of the treatments may influence the results. Different sequences of administration of the treatments can help to balance or minimize the influence of the order of administration.
- Sixth is whether a placebo or other control treatment is included. A placebo is a neutral or ineffective treatment given as if it were an effective treatment. A control treatment is usually the continuation of current practice. These placebo and control treatments reduce the bias of the expectations of subjects in the study. An ethical dilemma exists when an anticipated effective treatment is not given and the persons receive an ineffective placebo or an ineffective control treatment.
- Seventh is the inclusion of multiple dependent variables and measures. Multivariate means using multiple dependent variables or measures to provide greater understanding of the effects of the independent variable(s). Significant effects of the independent variable(s) may only show in one dependent variable or measure. In that case, the researcher has limited ability to generalize conclusions.

- Eighth is latency of time. A longer period may be required to detect the effects of manipulation of an independent variable. Delayed effects are common in many economic policies and psychological therapy.
- Ninth is unique to single-subject research. This research first requires establishing a stable baseline of a frequency or intensity of a behavior. Determining the success of the treatment requires a pattern of administering and removing the treatment. Sufficient time must elapse between administration and removal to determine the lasting and latent effects of the treatment. The results may not generalize to other persons.

Department of Education standards for rigorous research include three types of broad research designs. The first encompasses different forms of randomized controlled groups design, which randomly assigns persons to different groups. A control group(s) or placebo treatment is a neutral comparison. A second design is regression-discontinuity, which uses a cut point on a variable, rather than random assignment, to assign individuals to establish comparability between groups. The third design is single-subject case studies, in particular applications for intervention studies after establishment of a reliable baseline.

Many specific experimental research designs have been proposed. Each specific design is most effective to control for certain types of potential threats to the validity of a study. The particular research question and practical constraints indicate the most appropriate experimental research design. A complete listing of specific experimental designs is also beyond the scope of this introductory textbook.

These designs incorporate many different techniques to try to minimize the threats to the validity of the research study. The techniques vary in effectiveness in terms of controlling these threats and reducing errors in gathering and interpreting experimental research information. The following is a list of different techniques to minimize threats to validity in experimental research:

Techniques to Minimize Threats to Validity in Research Designs

- Randomly assign persons to different groups receiving different treatments.
- Match individuals on a characteristic and then randomly assign each to different groups, so that the different groups have the same number of persons with that characteristic. This has no assurance of equality with other characteristics.
- Consider subject characteristics as a constant, where each person in the study has the same type or amount of a characteristic.

- Categorize each person in a subgroup or range of values on a characteristic and then randomly assign persons so that each group has the same proportion of persons in that subgroup.
- Repeat treatment such that each group receives each treatment, so that there is no difference between groups.
- Adjust scores based upon each person's score on another variable or pretest.
- Utilize a single-blind design, wherein the subjects are not knowledgeable about the treatment.
- Utilize a double-blind design, wherein neither the subjects nor the experimenter are knowledgeable about the treatment.
- Consider subject characteristics as independent variables.
- Establish a control group that does not receive a placebo treatment and continues current practice.
- Use placebo treatment to address attitudes and expectations of persons in the study.
- Use multiple experimenters to gather the information.
- Ensure treatment fidelity through monitoring of administration of the treatments so that each person or group receives the treatment as intended.
- Use more reliable and valid data collection instruments.
- Use multiple data collection instruments and multiple types of instruments.

Three other types of research are frequently used to gather information:

1. In action research, practitioners collect current data as a guide to improve the current delivery of a treatment. This type of research is used by practitioners with a focus on the local situation rather than generalization to other situations. This research has limitations in controlling other potential explanations.
2. In ethnographic research, there is a description of a culture or a cultural practice. Data are collected from observations of groups or individuals. This research has many uncontrolled potential explanations that limit the conclusions.
3. In mixed-methods research, experimental and non-experimental research with surveys and interviews combine to provide more explanation for obtained results than any separate method. Agreement from each method gives greater confidence in the accuracy of the results.

COMMON ERRORS

1. No equivalent groups used as comparison groups

2. No equivalent control group(s) used
3. Comparison group(s) significantly disproportionate in size or composition of subject characteristics
4. Disproportionate number of missing data for comparison groups
5. All possible combinations of groups not included in factorial design
6. All sequences of treatments not included in counterbalanced design
7. Insufficient number of observations either before or after an intervention in a time series design with multiple times of data collection
8. Nonsignificant characteristic and/or imprecise matching characteristic used for matching
9. Inaccurate assumption of fixed (only one type or amount) or random (range of types or amounts) independent variable
10. Inaccurately addressing a nested design in which results are restricted to a single level of another independent variable
11. Short length of time between administration and second administration of treatment in a single-subject research study (does not permit demonstration of treatment effects)
12. Excessive amount of time between administration and second administration of a treatment in a single-subject research study (permits the influence of additional factors to minimize the effects of the treatment)
13. Insufficient detail provided to replicate the study/detail (indicates inadequate standardization of procedures for different groups of persons)

EXAMPLES

- *The Plain Dealer* reported that a study funded by the Templeton Foundation found no difference in recovery from surgery for those prayed for and those not prayed for. There were three groups: (a) prayed for and knew it, (b) prayed for but only knew it was a possibility, and (c) not prayed for but told it was a possibility. There was no group not prayed for and not told it was possible. This created the error of a missing equivalent control group, as not all combinations of the independent variables of prayer and knowledge of prayer were included.
- Robert Blum of the Johns Hopkins University reported, in the journal *Contraception*, a review of many studies involving more than 150,000 women. Studies with flawed methodology involving inappropriate control groups, unreliable mental health measures, and inappropriate sampling methods indicated that women who had abortions have lower long-term mental health than women who did not have abortions. Studies with more appropriate methodology involving control groups, measures, and sam-

pling found no significant difference between women who did and did not have an abortion. Errors in methodology can contribute to an erroneous conclusion. Despite this evidence, the United States Supreme Court still cites discredited and erroneous research as a rationale for limiting late-term abortions. (Should justices accurately interpret research information? Should laws reflect accurate research evidence?)

- *The Plain Dealer* summarized a study by David De Vaus of La Trobe University in Melbourne that found no difference between men and women in terms of the frequency of emotional problems or happiness in marriage. These results were based upon interviews with 10,641 persons and contrasted with a long-standing belief postulated by Jessie Bernard in 1972 that marriage makes men happier and women more miserable because marriage oppresses women. Bernard's research was flawed and erroneous because of the use of unreliable measures of happiness and a lack of control for equivalent subject characteristics. Bernard's discredited theory is still widely accepted and has enormous consequences, erroneously influencing judicial decisions. (By the way, married men and women have the same rate of mental disorders. However, women have more anxiety and depression and men have more alcohol and drug abuse.)
- Research by Retina Foundation of the Southwest, reported in *The Plain Dealer*, indicated that adding nutrients makes bottled infant formula just as effective as breast milk for the mental development of infants. An error was not having an equivalent control group of babies who were breast-fed so that the research is unable to support the conclusion.
- *The New York Times* reported that Dr. Werner Bezwoda admitted that he falsified research data claiming that bone marrow transplants and high-dose chemotherapy increase longevity in women with breast cancer. An inaccurate control group received a treatment other than the conventional treatment cited in the findings. Is there any harm in having patients go through an arduous, expensive, and worthless treatment?

REDUCING ERRORS

An appropriate research design can reduce errors by providing guidance for selecting methods to gather and interpret information. Misalignment of the research design with the research question or errors in implementing the various procedures described throughout this textbook can contribute to erroneous conclusions.

Reduction of errors occurs in varying degrees with different research designs and procedures. Assignment of individuals to comparison groups by randomization or in a manner that minimizes any initial differences among groups reduces errors. Persons can also be matched on a characteristic or stratum of interest and then randomly assigned equally to the different comparison groups. Persons who vary slightly on a characteristic or stratum are homogeneous on that characteristic and can be randomly assigned to different groups, but with slightly less assurance of the initial equality of the groups.

Other procedures may be more limited in their ability in reduce errors. One procedure is to have every person receive each manipulation or treatment applied in a research study. Concerns are the order in which the treatments are given and the residual or carryover effect. Another procedure involves ensuring that each person in the study has the same or a constant type or degree of a characteristic. The limited generalizability of results to other types of persons is a concern with this method.

A statistical method—analysis of covariance—statistically equates and minimizes differences among comparison groups by using another variable to adjust scores. A covariate is the variable that is the basis for making the statistical adjustment. The appropriateness of the covariate is a concern with this procedure.

It is common that participants do not accurately implement what they are required to do. Variation in the accuracy of implementation is a significant contributor to errors in research. Monitoring the research study—to verify that participants are accurately performing or implementing the required tasks—is important to reduce errors and increase treatment fidelity.

BEST PRACTICES

- Select the type of research design that aligns with the type of research question (e.g., descriptive, comparative, correlational, predictive, causal-comparative, ex post facto, quasi-experimental, experimental, single subject, action, ethnographic, mixed-method).
- Select a specific research design that aligns with the important considerations described in this chapter (e.g., number of subjects, type or amount of the independent variable(s), number of independent variables, pretesting, number of treatment combinations administered to each subject, placebo, number of dependent variables, latency of time measuring effects, stability of baseline).
- Select the specific research techniques to minimize threats to internal and external validity (randomization, matching, subject characteristics as a constant, subgroups, repeated treatment, covariance analysis, single-blind design, double-blind design, subject characteristic as a variable, more valid and reliable instruments, multiple instruments, placebo group, control group, treatment fidelity with continuous monitoring of treatment implementation).

- Utilize different types of randomized control group designs for a high degree of control in experimental research where a control or placebo group is used.
- Utilize different types of regression discontinuity analysis designs for a high degree of control in experimental research where a control or placebo group is not used.
- Replicate original research to provide greater confidence in results.
- Use different research designs or mixed-methods research to provide greater confidence in and generalizability of results.
- Use a single design with multiple independent variables (factorial) rather than multiple studies that use only a single independent variable to obtain interaction and reduce errors from conducting multiple studies.
- Use a single study with multiple (multivariate) dependent variables rather than multiple studies that use only a single dependent variable to provide a more comprehensive understanding of effects of treatment and reduce errors from multiple studies.

INTERRELATION TO OTHER TOPICS

The research design provides an overall plan for establishing the research activities. The next chapter provides information about the skills needed and the errors made in identifying important characteristics in the obtained evidence. These characteristics help us to select appropriate statistical methods to analyze and interpret the evidence.

Chapter Ten

Data Analysis

"Statistics Whisperer"

Not all data are created equal.

—Anonymous (we really don't know who said this)

Now go, write it before them in a table, and note it in a book.

—The Holy Bible

One picture is worth a thousand words.

—Anonymous

Given for one instant an intelligence which could comprehend all the forces by which nature is animated and . . . submit these data to analysis, . . . nothing would be uncertain.

—Pierre Laplace

Statistical figures referring to economic events are historical data. They tell us what happened in a non-repeatable historical case.

—Ludwig von Mises

Wall Street indexes predicted nine out of the last five recessions.

—Paul Samuelson

SIGNIFICANCE

A commonly held assumption is that statistics can show any conclusion. The real world supports this assumption because of many situations that select inappropriate statistical procedures or, regrettably, use fabricated or falsified information.

Many erroneous decisions are the result of inaccurate or inappropriate statistical analysis procedures. Selecting the appropriate statistical procedure requires an understanding of the type of research question and the characteristics of the data. This chapter provides a description of the multiple types of characteristics of data to guide the selection of appropriate statistical procedures.

Contrary to common belief, usually only one or two different statistical procedures are appropriate to analyze the obtained information. The number of decimal places to which a statistical procedure is calculated may be irrelevant if the procedure is incorrect for the obtained data. Making accurate decisions requires the use of appropriate statistical procedures.

This chapter includes different methods of presenting and analyzing descriptive statistics, such as tables, graphs, measures of central tendency, and standard scores, as well as limitations of percentiles and age or grade equivalents. Representative statistical procedures are included for answering correlational, prediction, or cause-effect questions.

LEARNING OBJECTIVES

1. Understand that data have different characteristics considered in selecting the appropriate method of statistical analysis.
2. Understand the concept of a normal distribution.
3. Understand the concept of linear and nonlinear distribution of scores.
4. Understand the concept of homogeneity of variance.
5. Understand that different statistical analysis methods assume and require that data have certain characteristics.
6. Understand the different types of graphical presentation of data.
7. Understand concerns that affect the accurate graphical presentation of data.
8. Understand the different types of tables.
9. Understand the concept of grouping scores into categories.
10. Understand the advantages and limitations of grouping scores.
11. Understand the concept of skewed distribution.
12. Understand the different measures of central tendency (mode, median, mean).
13. Understand the concept of range and standard deviation.
14. Understand the concept of percentile and the advantages and disadvantages of percentiles.
15. Understand the limitations of age-equivalent and grade-equivalent scores.

16. Understand the concept of standard score and the advantages and disadvantages of standard scores.
17. Understand the commonly used standard scores.
18. Understand what the correlation coefficient indicates about a relationship.
19. Understand the concept of regression and multiple regression in prediction.
20. Understand the concept of eta in identifying the degree of curvilinearity in a relationship.
21. Understand the concept of effect size in identifying significant differences.
22. Understand the concept of standard error of estimate in prediction.
23. Understand the concept of standard error of the mean.
24. Understand the concept of standard error of difference in comparing groups.
25. Understand the concept of t-tests.
26. Understand the concept of analysis of variance.
27. Understand the different forms of analysis of variance.
28. Understand the advantage of analysis of variance compared to t-tests.
29. Understand the concept of nonparametric statistical tests.
30. Understand the advantage of parametric statistical tests compared to nonparametric statistical tests.
31. Understand different types of nonparametric statistical tests.

PREREQUISITES

1. Understand the foundations of research instrument measurement.
2. Understand the different types of research questions.
3. Understand the concept of dependent variable.
4. Understand the concept of operational definition.

CONCEPTS

We collect so much information in our daily lives. This information helps us to describe, summarize, identify relationships, and infer causes. Identifying and understanding important characteristics and details of the data helps us to accurately process this information in making decisions. Individuals and a society need to be able to distinguish between inaccurate and accurate information and between inaccurate and accurate statistical analysis.

Researchers also need to identify characteristics of data to select the most appropriate statistical procedures. Previous chapters discussed validity, reliability, levels of measurement, type of research question, norm- and criterion-referenced scores, and the threats to the validity of the research study that can contribute to inaccurate information.

This chapter describes additional characteristics of data to guide the selection of the appropriate method of statistical analysis. The first is a normal distribution of scores, in which most people and scores cluster around a central value, with fewer people and scores either way above or below that score. The distribution of scores is symmetrical, with an equal number of persons or scores above and below the central value. Normal distribution of scores is an important assumption in the class of statistical analysis called parametric statistical analysis.

Most characteristics have a normal distribution, such as height, weight, and even intelligence. An important consideration is the norm or comparison group. A normal distribution of scores may occur within one group, but not within another group. What is "normal" in one group may not be "normal" in another. A man that is 7 feet tall is not of normal height in a group of horse jockeys, but his height would be normal in a group of basketball centers.

A skewed distribution has scores asymmetrically distributed around a central point. A positive skewed distribution has very few high scores relative to low scores. A negative skewed distribution has very few low scores relative to high scores. Asymmetry of scores violates assumptions of many mathematical procedures. Significant mathematical errors result from using parametric statistical analysis methods with a skewed distribution of scores. Nonparametric statistical methods comprise a special class of statistical analysis for skewed distributions.

The second concept is homogeneity of variance, i.e., the equivalence of the spread or variability of scores or persons in one group and another group. The third concept, linearity, refers to the linear or nonlinear relationship among scores of different persons or groups, such that a linear relationship (positive or negative) continues in the same direction through the entire range of possible scores. An eta correlation coefficient measures the degree of a nonlinear or curvilinear relationship. A higher eta indicates that the relationship changes direction over the range of possible scores.

Mistakes in understanding the details of the data collected in an investigation contribute to selecting an inappropriate method of statistical analysis and an inaccurate interpretation of the results of the investigation. The following are important characteristics of the obtained data in the selection of the appropriate method of statistical analysis:

Characteristics of Data in Selecting Method of Statistical Analysis

- number of independent or predictor variables
- number of dependent variables
- number of groups being compared

- relationship between groups (independent if groups comprise different persons or dependent if groups include same persons or matched persons)
- number of persons in a group
- relative proportion of persons in comparison groups
- level of measurement
- type of research question
- normal or skewed distribution of scores
- number of observations or collected data from each person
- type of score measurement (norm-referenced with standard score or percentile, criterion-referenced with percentage, or frequency)
- linear or nonlinear relationship between scores
- homogeneity or equality of variability of scores in different groups
- unit of analysis of number of groups or number of persons in groups
- proportion of comparison groups with missing data

Information and data come into our daily lives in so many ways that we may feel under constant barrage, until we just want to shout, "Incoming!" The methods in which data are organized and presented can affect our interpretations and our decisions. Powerful visual images like graphs or deceptive phrases, such as "the average American" rather than "the average of Americans," can influence our decisions. Whether these methods of presenting data are intentionally or unintentionally inaccurate, the result is the same: an inaccurate decision.

Graphs can present a visual image of the distribution of scores. The two types of graphs generally used are the histogram and the frequency polygon. The histogram, or bar graph, uses bars whose height or length shows the frequency of a score. The frequency polygon has a series of dots indicating the frequency of the score. The overall distribution of scores is indicated when the dots are connected.

Common errors with these graphs include truncating or cutting off part of the range of distribution of the scores, having too small or too large a scale of distance between possible scores, and not indicating instances where there was no obtained score. The visual image can provide a distorted picture.

Several types of tables can be used to summarize data. The first is a frequency distribution, which shows the frequency of each obtained score. The second is a cumulative frequency distribution, which indicates the number of times a particular score or one of smaller value has already occurred.

Grouped data encompasses the presentation of information in graphs, tables, and for further statistical analysis. It reports the frequency with which a score appears in a category or a group of several consecutive scores, rather than reporting each individual score. Grouping data in graphs or tables helps to condense a great amount of data, since there are fewer categories than individual scores. Some accuracy of the data is lost because of the uncertainty of the frequency of each particular score within the category. Adjustments are required in statistical analysis of grouped data, with a common assumption that all scores within a category are at the midpoint of the category.

Another way to analyze the obtained data is with the calculation of a measure of central tendency, which provides one measure indicating a typical score. Three measures of central tendency are the mode, median, and mean. The mode is the score that appears most often. The possibility of having more than one mode is a concern. The median is the value above and below which an equal number of persons scored and is the 50th percentile. Extreme scores do not affect it. The mean (M) is the average and requires at least an interval level of measurement to calculate. For example, there can be an average age, but not an average eye color. Skewed data or a few extreme scores affect the mean.

In addition to trying to identify the central tendency of the scores, there is the need to examine the variability of scores. The two most often used measures are the range and the standard deviation (SD). The range is the difference between the highest and lowest obtained scores. The standard deviation is more complicated and indicates the average deviation of the scores from the mean or the average of the group.

The mean and the standard deviation are important concepts applied to many situations to indicate how well a person does compared to others. A type of score called a standard score indicates how many standard deviations a person is from the average of the group. These scores give a uniform basis of comparing persons and do not exaggerate small differences in the actual scores. The standard scores commonly used to understand and report norm-referenced data include the following:

Common Types of Standard Scores

- Deviation IQ has a mean of 50 and a standard deviation of 15; used to indicate a person's IQ.
- Normal curve equivalent (NCE) has a mean of 50 and a standard deviation of 21.06; used by the government to report results of federally funded programs.
- Z score has a mean of 0 and a standard deviation of 1; used in many statistical analyses that indicate the number of standard deviations from the average.
- T score has a mean of 50 and a standard deviation of 10; used in educational settings to indicate a level of achievement.
- Stanine has a mean of 5 and a standard deviation of 2; used in educational settings to indicate a level of achievement.

Some calculated scores are misinterpreted and used with caution. The percentile score indicates the percentage of others surpassed by an individual in a group. Percentile scores exaggerate small differences in actual performance for those in the middle of a distribution. Percentile scores are an ordinal-level measurement, and it is erroneous to calculate an average percentile score. Percentiles can convert into a standard score to permit calculation of an average.

Age or grade equivalent scores have four limitations that contribute to erroneous interpretation. The first is that lower age and grade equivalents on one measure sometimes represent better performance than a higher score on a different measure. The second is that age and grade equivalents are ordinal-level measurements that represent unequal amounts of performance from one score to the next. The third is the erroneous calculation of an average age or grade equivalent score. The fourth is that grade equivalents are estimated scores and do not indicate grade level mastery. In spite of these cautions, schools use grade equivalents in making decisions about students.

Sometimes the goal is to examine the relationship between changes in the scores of one variable and the scores of another variable. Correlation coefficients calculate both the direction of the relationship (positive if the scores of both variables change in the same direction and negative if the scores change in opposite directions) and the strength of the relationship (the more consistent the rate of change, the stronger the relationship). A very common and dangerous error is assuming that correlations indicate cause and effect.

Correlations allow us to predict a score on one variable based on knowledge of the score on another variable. The statistical techniques of regression or multiple regression incorporate correlations to predict the scores on a variable. Regression uses the scores on one variable to make the prediction, and multiple regression combines the scores from several predictor variables to make the prediction. Many models of regression analysis are available, and each makes different assumptions about the order and the procedure to combine the multiple predictors.

A standard error of estimate indicates the average amount of error in predicting a score. A small standard error of estimate indicates a smaller degree of error in prediction. We also have questions regarding significant differences between groups. Various methods, such as t-tests (two groups) or the different forms of analysis of variance or ANOVA (many groups), compare the averages of different groups. A standard error of the difference is the basis for determining if the difference among the averages of the groups was greater than expected. The effects of multiple independent variables (factorial) and multiple dependent variables (multivariate) are determined with different variations of the analysis of variance.

Nonparametric statistics is a category of different statistical methods for use whenever any of three characteristics of data are present. The first is a non-normal distribution of scores. The second is less than an interval level of measurement. The third is low homogeneity of variance where the scores in a group have great variation. Multiple nonparametric statistical procedures are available. Nonparametric statistical tests have less statistical power to reject a false null hypothesis and less ability to assess the practical significance of the results (effect size).

The goal of data analysis is to provide an accurate representation of the data collected as the basis for appropriate interpretation. Researchers are ethically required to keep their original data for several years after the conclusion of an investigation to permit other researchers to corroborate the accuracy of the data and the interpretations. A very recent example of the need for corroborating investigations is the detection of coding errors in entering data about the relationship between national debt and economic growth that supported an incorrect economic policy of austerity.

Unfortunately, a complete description of data analysis is not possible in this introductory textbook. Hundreds of statistical methods are available. A great way to fall asleep is to count the number of different methods. Here are some: the Mann-Whitney U test, the Kolmogorov-Smirnov goodness-of-fit test, the Wilcoxon matched-pairs signed-ranks test, the Bartlett test for equality of variance, and the Dunn-Bonferroni procedure. Are we still awake?

Other textbooks provide a more complete discussion of the advantages and disadvantages of the different methods to present the data. Those authors will have all the fun of seeing the joy and excitement that we gain from studying statistics.

COMMON ERRORS

1. Inaccurate number of independent or predictor variables
2. Inaccurate number of dependent variables
3. Inaccurate number of groups
4. Inaccurate relationships between groups
5. Inaccurate number of persons in the groups
6. Inaccurate relative number of persons in comparison groups
7. Inappropriate level of measurement
8. Inaccuracy in type of question investigated
9. Inaccurate assumption of normal distribution of scores
10. Inaccurate number of observations or collection of data from each person
11. Inaccurate type of score measurement (norm- or criterion-referenced)
12. Inaccurate assumption or lack of knowledge of linear relationship between scores

13. Homogeneity or equality of variability of scores in different groups unknown or inaccurate
14. Inappropriate unit of analysis accounting for number of groups rather than number of subjects
15. Inappropriate unit of analysis accounting for number of subjects rather than number of groups
16. Disproportionate and/or unknown missing data for different groups
17. Truncated distribution of scores in a graph or table such that description of data is distorted
18. Small scale of dimensions in graph such that description of data is distorted
19. Large scale of dimensions in graph such that description of data is distorted
20. Placement of beginning or ending points along dimensions in graph or table such that description of data is distorted
21. Insufficient number of categories along dimension in graph or table such that description of data is distorted
22. Missing data or no response for a particular value not indicated in graph or table
23. Small score intervals in graph or table such that description of data is distorted
24. Large score intervals in graph or table such that description of data is distorted
25. Use of pictograph that distorts description of data
26. Inappropriate method of statistical analysis used for:

    a. Number of independent variables
    b. Number of dependent variables
    c. Level of measurement
    d. Assumption of normality
    e. Relationship between groups
    f. Number of groups
    g. Type of research question
    h. Homogeneity of variance
    i. Assumption of linearity of relationship
    j. Number of persons in a group
    k. Number of observations for each person
    l. Unit of analysis as either number of persons or number of groups
    m. Proportion of missing data
    n. Use of covariates
    o. Number of covariates

27. Measure of variability not provided
28. Large standard error of estimate that adversely affects confidence in accuracy of predicted scores
29. Large standard error of the mean that adversely affects confidence in accuracy in obtained mean
30. Large standard error of the difference that adversely affects confidence in accuracy of difference between means of groups
31. Use of inappropriate correlation method based upon incorrect assumption of artificial or real dichotomous variables
32. Use of inappropriate correlation method because of truncated or skewed distribution of scores
33. Use of inappropriate regression method based upon lack of robustness to violations of data assumptions
34. Use of inappropriate regression method where coefficients of a hierarchical linear model are affected by other variables
35. The predicted criterion variable has low reliability
36. Large shrinkage in accuracy or anticipated shrinkage not indicated in predictions
37. Proportion of variance explained small or not indicated
38. Multi-collinearity large or not reported
39. Eta large or unreported
40. Effect size small or not reported
41. Statistical power low or not indicated
42. Lack of covariates or covariates resulting in insufficient degree of adjustment of scores when unsure of initial group differences
43. Use of parametric statistical tests rather than nonparametric statistical tests
44. Use of assumed or projected data rather than actual data
45. Use of inconsistent methods to collect data for comparisons
46. Average percentile calculated
47. Average age or grade equivalent calculated
48. Incorrect assumption of actual scores rather than grouped scores in calculations
49. Midpoint of grouped scores not used in calculations
50. Midpoint incorrectly used in a truncated category in calculations

EXAMPLES

- Timothy Noah, in *The Great Divergence*, reported that income growth in America since 1980 has not been distributed normally. Over 80% of the total increase has gone to the top 1% of persons. Calculations indicating average income growth are erroneous because income in America is not normally distributed.

- Dan Ariely reported, in *Predictably Irrational*, that wealth in America is not normally distributed. The bottom half of Americans has a very small percentage of the wealth, and the top half of Americans has all but a small percentage of the total wealth. American wealth is not normally distributed and thus it is erroneous to calculate average wealth in America.
- President Bush repeatedly reported that his tax reductions would give the average taxpayer a tax cut of $1,083. But this average is not what the typical taxpayer would get, as the top 1% would get $24,428. The error is that the average was calculated on a non-normal distribution that skews and exaggerates the averages.
- Michael Vick, NFL quarterback, was quoted in *The Plain Dealer* as saying, "I have two weapons—my legs, my arm, and my brains." (Doggone it, we will try to avoid calling out the hounds for this error in data analysis. We could be barking up the wrong tree, but Vick's math is not correct.)
- *The Plain Dealer* reported that Governor John Kasich of Ohio made multiple statistical errors in claiming 1,800 automotive job losses in Ohio because of the automotive bailout. First, he used a truncated time period, from when he took office in January 2011 until September 2011. This time period ignored the 27,000 Ohio autoworkers who went back to work during the first 27 months of the automotive bailout, from its beginning in June 2009 until September 2011. Second, he switched methods used to calculate jobs, selecting a less comprehensive Bureau of Labor Statistics data set that used estimates from a sample of employers. Data sets based upon actual jobs—counted by the Quarterly Census of Employment and Wages of the Bureau of Labor Statistics—provided a more accurate total of jobs that contradicted Kasich's figures. The errors involved the use of assumed data rather than actual data, inappropriate start and end times to determine employment, and inconsistent methods to collect data for comparison.
- Joe Rojas-Burke reported, in *The Plain Dealer*, that colonoscopy screenings for colon cancer detect 80%–95% of tumors at a cost of about $900, compared to sigmoidoscopy, a procedure that detects 60% of tumors at a cost of $150. Insurance companies are reluctant to pay for the colonoscopy because of the dilemma of determining whether the increased number of lives saved is worth the added cost. It is a cost-benefit analysis of the cost versus the reduction of errors in detection. What is the practical significance of increased accuracy of detection versus increased cost? Should a cost-benefit apply to us all? What are we worth?
- *The Plain Dealer* reported that the Cleveland Clinic developed a better predictor of heart attack or stroke by using the enzyme myeloperoxidase (MPO). MPO explains a greater percentage of the variance in heart attack

and stroke than the other predictors, high cholesterol and blood pressure. The use of an additional predictor increased the accuracy of the prediction and reduced the standard error of estimate in prediction.

- Nielsen ratings do not provide an exact indication of the age of television viewers, using categories such as those older than 65. The age distribution is truncated. Other ages in the Nielsen sample are presented as grouped data, such as ages 25–49. It would be an error to calculate an exact average age of viewers for a network based on this data. Age of the viewer is an important characteristic for determining advertising rates, so it would be better to indicate the percentage of viewers in each category.

REDUCING ERRORS

The public believes that statistics can support any conclusion researchers want. However, for any situation, the characteristics of the data indicate that only one or two different statistical procedures are appropriate. The characteristics of data described in this chapter provide a checklist to identify an appropriate procedure.

The important assumption is that the researcher or consumer of information is capable of accurately understanding each type of characteristic in the obtained data. An incorrect understanding of any of the characteristics in the data can contribute to errors through selection of an inappropriate procedure.

Selection of a procedure that is appropriate for the obtained data and can verify the accuracy of the mathematical calculations reduces statistical analysis errors. Miscoding or incorrectly entering data on a spreadsheet for analysis can contribute to erroneous conclusions.

Selecting the appropriate procedure and verifying the accuracy of the calculations are necessary steps. However, the prerequisite is that data are accurate.

Different types of scores and data reporting methods were included in this chapter. Many of these scores, such as percentiles, age or grade equivalents, averages, and correlation coefficients, are often misinterpreted.

Graphs can easily distort the information with truncated score dimensions, variations in beginning and ending points in a distribution, and variations in the size of scale intervals. Concurrent comparison with the data reported in tables can minimize errors in interpretation.

The following affect analysis of grouped data: (a) the size of the interval, (b) the starting points in the dimensions, (b) the truncation of upper or lower categories, and (d) the number of intervals. Grouped data assume that each person in the group or category received the score at the midpoint of that group or category. This assumption contributes to less precision and more errors in statistical analysis when using grouped as opposed to ungrouped scores.

BEST PRACTICES

- Align the characteristics of data described in this chapter to select a statistical method that is appropriate for the research question and the obtained data.
- Use interval or ratio level of measurement of data whenever possible for greater precision and power in statistical testing.
- Use standard scores rather than age and/or grade equivalent scores.
- Use ungrouped rather than grouped data whenever possible for greater precision because analysis is affected by size of interval, number of intervals, and starting and ending points of groups.
- Use percentile scores with caution, as these scores cannot be averaged.
- Avoid using data from truncated distributions of scores.
- Avoid using only a graph to present information, as graphs can easily create a distorted visual image.
- Use alternative analysis procedures whenever more than 15% of data are missing or when one group has fewer than two-thirds of the participants of other groups. Although no universally agreed upon method exists, the most common include analyzing only participants with complete data or substituting an average for the missing data.

INTERRELATION TO OTHER TOPICS

The characteristics of the information assist in selecting the appropriate statistical analysis for making decisions. An incorrect method of statistical analysis can contribute to erroneous decisions. As often mentioned in this textbook, research provides only tentative conclusions based upon probability. The next chapter also discusses practical significance.

Chapter Eleven

Statistical Probability and Interpretation

"The Wizard of Odds"

The theory of probabilities is at bottom nothing but common sense reduced to calculus.

—Pierre Laplace

As for a future life, every man must judge for himself between conflicting vague probabilities.

—Charles Darwin

There is no such thing as absolute certainty, but there is assurance sufficient for the purposes of human life.

—John Stuart Mill

The public . . . demands certainties. . . . But there are no certainties.

—Henry Mencken

Statistics means never having to say that you are certain.

—Anonymous (we think this is from us, but we are not sure)

Conscience is the perfect interpreter of life.

—Karl Barth

Man is the interpreter of nature, science the right interpretation.

—William Whewell

SIGNIFICANCE

As the old joke goes, nothing in life is certain besides death and taxes. A recurring theme of this textbook is the tentativeness of research results. Reducing errors can provide accurate knowledge about the probability of the results and the limitations of knowledge. Not appropriately considering the probability of an event or the significance of an unanticipated event causes serious errors. These errors may involve underestimating or overestimating the probability of an event. These inaccurate estimates may reflect the use of inappropriate research and statistical analysis procedures or our own inability to accept the probability of the results.

Overconfidence is a significant contributor to inaccurately considering both the probability of a result and the consequences of an adverse outcome. Inaccurately estimating the practical significance of either beneficial or adverse unanticipated results can have profound consequences for our decisions.

Researchers conduct controlled studies to determine the probability of an event, but are not always able to determine the practical significance of an erroneous decision. Errors can have a low probability of occurrence, but the impact of errors may have significant effects. Statistical probability is only a part of the overall consideration in using information to make decisions. Research only provides tentative results based upon probability. Any interpretation must also ask the question: What are the consequences if we are in error?

This chapter provides some examples of errors involving inappropriate generalizations, inappropriate research procedures and statistical analysis, and ignorance of the consequences of potential erroneous outcomes. Misinterpretation can be the result of multiple potential errors.

LEARNING OBJECTIVES

1. Understand that research provides only tentative conclusions based upon probability rather than certainty.
2. Understand that probability only refers to the result as not due to chance or other factors.
3. Understand that statistical significance only refers to probability and not to the practical significance of the result.
4. Understand that a symbol ($p < .01$) of statistical probability indicates only the probability of a Type 1 error.
5. Understand that probability does not refer to the importance of the consequence of an erroneous result.

6. Understand that unanticipated results should be reported as such rather than as an ex post facto hypothesis.
7. Understand that interpretation includes the reporting of statistical power and effect size.

PREREQUISITES

1. Understand the data presentation and analysis methods.
2. Understand the threats to internal and external validity of a research study.
3. Understand the type of information provided by different data analysis methods.
4. Understand the concept of statistical probability.
5. Understand the concepts of Type 1 and Type 2 errors.

CONCEPTS

We have no absolute certainty in many of the actions that we undertake in everyday living. The concept of uncertainty or probability is often referred to in common language, but many of us have difficulty understanding the concept of probability in making decisions.

Phrases we hear about probability include: "Certain beyond a reasonable doubt"; "Probable cause"; "I'll take my chances"; "What are the odds?"; "Chances are slim"; "Fat chance"; "Don't bet on it"; "Against all odds"; "The deck is stacked against you"; "The market hates uncertainty"; "I feel lucky and just know that I'm going to win the lottery"; "Should I take a chance and introduce myself?"

Here are some additional phrases about probability: "Trust me, it's a slam dunk"; "Oddsmakers made the team a four-point favorite"; "Let's roll the dice"; "It's a black swan event"; "You're one in a million"; "Scarcity creates value"; "It's a miracle"; "Risk tolerance"; "It will never happen"; "Circumstantial or indisputable evidence."

No matter the importance of the decision, we all live with uncertainty in the decisions that we make. Do people overvalue certainty? Will someone stay in an abusive relationship that is certain instead of entering into a new relationship that is uncertain? Do we seek beliefs that promise the certainty of an eternal afterlife?

The common understanding of probability is the chance that an event will take place. The purchase of a life insurance policy is an example of probability. Insurance underwriters perform statistical analyses to determine the probability or the chances that we will live for a certain number of additional years. Information regarding our health history and health habits determines the amount of the life insurance premium that we pay. We are making a bet with the life insurance company, in which the life insurance company bets that we will live a long time, while we pay the life insurance premium and bet that we will not live a long time. (We are dead serious. That is the bet that the insurance company makes with us.)

Pascal's Wager is a classic dilemma in decision making that examines the different possibilities regarding whether God does or does not exist. Here are the possibilities:

1. God exists and we believe that God exists.
2. God does not exist and we believe that God exists. (Error)
3. God exists and we believe that God does not exist. (Error)
4. God does not exist and we believe that God does not exist.

Which of these errors should we take the most care to avoid? Can we calculate probabilities, and are the consequences of the errors more important than probabilities?

What is the probability of a miracle? Are miracles just errors because of the law of large numbers, which states that an event with a low probability in a small number of trials has a high probability in a large number of trials? An occurrence that is one in a million still can happen 310 times, because there are 310 million people in the United States. Is this a miracle?

Here is a situation to contemplate. We have a cognitive bias, as described in a later chapter, to worry about the improbable and vivid event but ignore the highly probable and not so vivid event. We worry more about an airplane crashing because it is a vivid but less probable event than death by the more probable event of slipping in a shower.

Researchers have a more specific meaning of probability: the likelihood that the obtained results are due to chance, errors in procedures, or an explanation other than the research hypothesis. Every statistical interpretation of research results must acknowledge each of these possibilities. Statistical probability uses a symbol ($p < .01$) to indicate the probability of a Type 1 error. In this example, the probability is less than 1%.

Researchers also have to acknowledge a degree of uncertainty in the results of an investigation by indicating that research never proves anything. The results of any investigation are tentative. Further research may disconfirm results of previous research. Variables or causes yet unknown can explain the results of the investigation. The search for these unknown, but potentially more appropriate, explanations is a powerful motivating force for continued research in all fields of knowledge.

We all collect information and hope to gain insight into the meaning of our lives and the events around us. This information can help guide us in making better future choices as individuals and as a society. Misinterpretation of the comments or behaviors of others can contribute to making less-appropriate future choices. Have we ever been in a situation in which we said something, but another person misinterpreted what we said? Did that create a problem?

Statistical probability or significance only suggests the likelihood of chance, errors, or other explanations. Understanding the statistical probability of Type 1 and Type 2 errors only provides one way to interpret the information. Interpretation should include the practical significance or meaning of the consequences of an incorrect conclusion. Individual and societal values may assist in determining the utility of the various correct or incorrect conclusions.

Researchers also have to be careful to make a correct interpretation of the results of an investigation. Correct interpretations can suggest appropriate conclusions and actions. Misinterpretations can suggest conclusions or actions that could have harmful consequences for current and future practice. Potential limitations or other explanations should be included when reporting research results and should be part of each individual's interpretation of the information.

Researchers need to report how the procedures and analyses used support the conclusion. As individuals, we also need to examine all available information and understand how we interpret the information.

COMMON ERRORS

1. Calculating inaccurate probability table value
2. Using incorrect probability table value where one-tail probability is assumed rather than two-tail as stated in hypothesis
3. Using large sample size that results in statistical significance
4. Not indicating practical significance
5. Using an assumed probability value rather than a calculated value
6. Not indicating probability of Type 1 and Type 2 errors
7. Not indicating practical consequence of Type 1 and Type 2 errors
8. Using statistical calculations with mathematical errors
9. Not indicating potential limitations in methodology, sampling, or instruments
10. Not indicating how a conclusion is supported by data
11. "Cherry picking" or using a part of data without complete indication of context or conditions
12. Not indicating methodology used to obtain information for conclusion
13. Claiming absolute basis of proof for a conclusion and ignoring probability

14. Inappropriately generalizing a conclusion from a group to all individuals in that group or a subgroup
15. Inappropriately generalizing a conclusion from a few individuals to a larger group
16. Inappropriately generalizing a conclusion to other:

    a. Types of independent variables
    b. Amounts of independent variables
    c. Number of independent variables
    d. Types of dependent variables
    e. Number of dependent variables
    f. Measures of the dependent variables
    g. Subject characteristics
    h. Conditions or locations
    i. Implementation procedures
    j. Covariates used in statistical analysis
    k. Time frames

17. Inappropriately generalizing a result from a small-scale study to a larger-scale implementation
18. Making a post hoc hypothesis and reporting results assuming an a priori hypothesis
19. Not indicating possibility of multiple causes and/or multiple effects
20. Logical inconsistency where same causes are not applied to similar effects (cuts in military spending create job loss but cuts in domestic spending do not create job loss?)
21. Not considering probability when any possibility is present
22. Assuming that a hypothetical calculation is an actual calculation
23. Interpreting a multi-collinear or confounding variable as the cause
24. Using a less precise level of measurement such as nominal (label) that distorts results
25. Inappropriately combining categories in grouped or categorical data

EXAMPLES

- The *Titanic* is unsinkable. This says it all regarding the lack of consideration for the probability and practical significance of errors. (We just thought that we would float this example by everyone. It is a “slam dunk” that we understand. Did we just mix metaphors?)
- Garth Sundem and John Tierney developed and refined a statistical formula that predicted the probability of a celebrity marriage lasting. The formula was published in *The New York Times* as part of the Sundem/Tierney

Unified Celebrity Theory, which makes a statistically significant probability estimate of divorce or an enduring marriage. The formula has made some correct predictions, along with some errors. What is the practical significance of this research? (We know what everyone is thinking. This textbook finally provides an example of a significant issue! We keep trying to find significant issues but have only found unimportant issues, such as world peace, errors resulting in deaths, and political manipulation.)

- The Bureau of the Census indicates that 47% of first marriages end in divorce. Of divorced persons, 75% remarry, but then 60% of the remarried divorce. Is this an error estimating chances of a happy marriage? (We wish we paid closer attention to combinations and permutations in mathematics so we could better calculate the odds of a happy marriage.)
- The Department of the Interior needed an accurate count of grizzly bears to determine whether they should be removed from the list of endangered species. A hypothetical estimate used reproductive and survival capacity rather than a count based upon actual observation. This hypothetical estimate has a greater potential for error compared to actual field observations. (The error could contribute to the elimination of the grizzly bear, and that result would truly be unbearable!)
- Christina Hoff Sommers summarized, in *The Atlantic*, the errors in methodology and the erroneous conclusion that schools shortchange girls as reported by the American Association of University Women. Those reports advocated for policies to correct the presumed disadvantages for girls. However, the reports had numerous errors, including missing data, and were not submitted to peer-reviewed journals for review.

  In contrast, many statistics from the Department of Education suggest that it is boys, and not girls, who are at a disadvantage, as boys have significantly lower rates of college admission, higher school dropout rates, and higher numbers in special education classes. The potentially erroneous conclusion that girls are at a disadvantage has led many professionals to dismiss any evidence indicating the possibility that it is boys who are at a disadvantage.
- Lenore Weitzman, in *The Divorce Revolution*, reported inaccurate data showing a 73% decline in the standard of living in women's households in the first year of divorce, compared to a 42% increase in men's households. Over 300 journal articles and 24 appeals to the United States Supreme Court cited the data as a centerpiece for changing America's divorce and child support laws. The problem is that the data were erroneously calculated.

  Weitzman did not make data from the report available to other researchers for several years. The data were then correctly recalculated and found a less significant disparity in the standard of living in the first year after divorce. Weitzman eventually admitted the data were in error, but the erroneous data had already affected legislation and policy decisions. The reported results contained mathematical errors.

- *The New York Times* reported that a Gallup poll on abortion was misinterpreted by columnist Ross Douthat, who "cherry picked" the data to state that 58% oppose abortion. The error was in combining the 20% who oppose abortion in all circumstances with the 38% who say that abortion should be legal only in some circumstances. The most relevant number is the 77% who stated that abortion should be legal in all or some circumstances. The error is the inappropriate combination of categories in categorical data.
- Gallup and other polls have indicated that only 40% of Americans say they are conservative, 35% say they are moderate, and 21% report that they are liberal. A repeated error is that most Americans are conservative. This is an error in interpretation of the difference between "most" and "more" in collapsing categorical data.
- *The Plain Dealer* reported that Ukrainian astronomers said that an asteroid 1,345 feet in diameter, 2013 TV135, could have a deadly collision with the earth on August 26, 2032. The odds are 1 in 63,000. What is the practical significance of the asteroid colliding with the earth? (Is this a low-probability, high-impact possibility error?)

REDUCING ERRORS

Reducing errors depends, in part, upon the accuracy of implementing appropriate research procedures as described in different chapters of this textbook. Errors in any of those procedures can produce erroneous results.

It may not be possible to compensate for errors in the implementation of any one of the procedures. The accuracy of research is only as good as the weakest link in the chain or sequence of research procedures.

Interpreting the research information in terms of the probability of occurrence and the practical significance of the consequences is an important task. Reducing errors in interpretation involves several steps. The first requires diligence in examining all procedures used in the process to gather the information. Errors in any of those procedures place constraints upon the probability of obtaining accurate results. Conducting appropriate statistical analysis and interpretation assumes that accurate information is gathered.

The second step is to review the accuracy of the statistical analysis procedures. The analysis should provide an estimate of the probability of the results due to chance. Statistical significance is only one factor in interpreting the results.

The third step is to examine the practical significance of the results. The importance or the meaning of the results is an important but often neglected component in interpreting results. The analysis should also include an estimate of statistical power and effect size as further guides to interpreting the practical significance of the results.

Statistical analysis can guide interpretation of probability and the degree of practical or meaningful differences. The fourth step in interpretation involves a value judgment of the utility of both a correct decision and the cost or consequence of an incorrect decision. This decision is often personal or societal and always involves both an element of uncertainty and the degree of importance attached to whether it is correct or incorrect.

Reducing errors in research procedures to prevent or minimize misinterpretation of statistical or practical significance in results is the focus of this textbook. Determining the values regarding the importance of the consequences of different outcomes or conclusions is another tool used to provide a more comprehensive interpretation of the results.

Statistics are numbers, but they do not always indicate the human cost to individuals who make up the statistics. Errors in statistical analysis have inevitable consequences that affect any decision.

BEST PRACTICES

- Examine practical significance of results in addition to statistical significance.
- Require a statistical effect size of equal to or greater than .25 of a standard deviation as significant.
- Require statistical power to reject a false null hypothesis equal to or greater than .90 if a Type 2 error is important.
- In calculations, use total number of subjects if the subjects independently experience the treatments and use one subject for each treatment if all subjects receive a treatment together.
- Use caution when interpreting statistically significant results with large numbers of subjects (more than 50 in a treatment group).
- The standard error of estimate in prediction should be equal to or less than 4% from a mean predicted score.
- Consider shrinkage or reduction in prediction accuracy in generalizing to a larger data set or another context.
- Use a coefficient of determination for estimating proportion of shared variance between variables in correlational research.
- Calculate the eta coefficient to linearity in the relationship among variables in correlational research.

- Use nonparametric tests when uncertain about meeting the required characteristics of data for parametric tests.

INTERRELATION TO OTHER TOPICS

Scientific research deals in probability rather than certainty. This chapter presented information regarding the statistical probability or the chances of obtaining certain results. An important concept is not only the probability of an event occurring, but also the practical significance and consequence of making an error that needs to be included when making a decision.

Perhaps the most important component in the sequence of procedures is our own biases and cognitive errors in interpreting the information. An exhaustive list of these errors is included in the next chapter. How many of these biases and cognitive errors have we had? How have they affected our beliefs and decisions? Will we be able to change?

Chapter Twelve

Personal Biases in Interpretation

"Gullible's Travels"

Decisions in life often are more the result of unconscious factors than conscious reasoning.

—Carl Jung

All things are ready, if our mind be so.

—Shakespeare

SIGNIFICANCE

Are we cowards, or do we have courage? Are we confident that we know what is right? These are provocative questions that each of us must answer.

We all have an image of who we are and images of others. Do we have accurate self-perceptions? Do we know why we believe and behave in certain ways? Do we know why others believe and behave in the ways that they do? Who we are and the cognitive processes or habits that we use can provide a filter through which to gather and then interpret information. Significant errors can be the result of inappropriate research procedures, but also our own distortions and biases.

Inappropriate dispositions and inaccurate cognitive processes can create a toxic combination contributing to many of our erroneous decisions. This textbook, unlike other textbooks about research methods, considers the biases and cognitive processes of an individual or a society as another aspect of research that is capable of producing erroneous decisions. Understanding how we process information becomes as important in reducing errors as selecting and accurately implementing the other research procedures described in this textbook.

An almost countless number of specific biases and cognitive distortions exist within each of us and are included in this chapter. These errors often reflect three broader and more fundamentally inaccurate assumptions within us. The first of these assumptions involves interpreting the information with a bias to confirm what we already know about ourselves or others rather than interpreting the information as a guide to change our beliefs or behaviors. This remains one of the most pervasive and significant errors in interpreting information.

A second fundamental inaccurate assumption regards the concept of randomness compared to purpose. It is human nature to be uncomfortable with ambiguous situations. We seek patterns and overgeneralizations to provide a suspected cause for an event. The consequence is that we overvalue apparent certainty and dismiss tentativeness or probability in information or conclusions.

A third fundamental inaccurate assumption involves the importance that we ascribe to social influences and pressure. We often feel uncomfortable with beliefs or behaviors that are inconsistent with those of others. We will seek confirmation from groups and become more reluctant to examine the possibility of the validity of other beliefs or conclusions.

The Romans constructed Hadrian's Wall in England to protect from invading tribes. The wall also marked the end of expansion of the Roman Empire. Do we fail to expand our knowledge and simply hide behind our own intellectual walls of bias and inaccurate thinking to protect ourselves from potentially contradictory information? The multiple types of biases and cognitive errors included in this chapter limit the expansion of knowledge. What is your intellectual Hadrian's Wall?

LEARNING OBJECTIVES

1. Understand that research principles apply to making personal decisions.
2. Understand that erroneous personal decisions may have significant implications for others.

PREREQUISITES

1. Be disposed to change beliefs when facts change.
2. Understand the concept that errors in research data also apply to personal decision making.
3. Understand the concept of probability.

CONCEPTS

The Declaration of Independence includes the idea that all of us have the unalienable right to life, liberty, and the pursuit of happiness. The liberty to make choices for the sake of our happiness is a gift, wrapped with the responsibility to ourselves and to others to consciously apply the reasoning of research methods to understand the available information.

Ben Jonson observed that true happiness in life comes from the worth of our choices. The hope is that each of us will make all future choices that will bring us happiness. How can we make such good choices? A prevailing notion still exists in some academic fields that people are rational in making decisions. That notion is increasingly suspect, as we all respond emotionally at times and do not always use higher reasoning and cognitive skills.

What does the lack of rationality do to the assumptions of the academic field of economics? Successful commercials often depend upon the assumption that people are not rational and that we have cognitive biases affecting our decisions. We purchase items that promise to make us healthier, richer, and more socially acceptable.

Listen to these sales pitches: "Just send the payment to us, but quickly, as there is a limited supply." "Hurry, a free car will be included for the first 50 people who send payment. In addition, a date with a movie star is included for the first 10 people who send payment."

The sad observation is that many will believe what is in that paragraph. Are people rational?

A growing field in behavioral economics indicates that we often superficially process information based upon cognitive and emotional errors. We later regret these choices. Are there 50 shades of rationality? Is there a clear answer to that question, or is that a gray area?

Researchers try to use best practices of research methods to understand the events and possibilities in many different academic fields. Researchers, in their personal lives, are also not immune to the unconscious factors and biases that are in all of us. Following best practices and taking precautions to guard against the multitude of possible errors should become part of the conscious reasoning in our lives, for the happiness of us all.

The following section lists common types of cognitive errors that affect decisions by distorting perceptions, assumptions, and reconstruction of events. How many of these cognitive errors do we make? Can we order a six-pack of biases to go?

Are there serious consequences of using these erroneous methods of interpreting information? Why do we use them? How will we change the way in which we interpret information?

COMMON ERRORS

1. Overgeneralizing results from a single or a few events to all other potential persons or events
2. Overgeneralizing results from a group or large number of events to all persons and/or events
3. Using an anecdote or personal experience rather than data
4. Using a less probable but more vivid event rather than a more probable but less vivid event
5. Publication bias that favors studies showing positive outcomes
6. Relying upon the most recent information
7. Relying on the first impression (primacy effect)
8. Relying upon faulty data analysis and interpretation of central tendency measures such as averages
9. Relying upon faulty data analysis of graphs
10. Relying upon faulty interpretation of correlation as causation
11. Relying only upon available information
12. Making decisions that reflect ideology and persevering in belief even when presented with contradictory data
13. Making decisions that conform to social pressure and holding beliefs of others to avoid conflict
14. Considering immediate effects rather than latent effects over time
15. Creation of bias through language or semantic manipulation

    a. Pareidolic bias: perception of a vague term or stimulus as significant (e.g., bureaucracy or seeing Jesus in a cloud)
    b. Pejorative bias: term is given negative meaning (e.g., big government)
    c. Hypocrisy bias: term is given a euphemistic positive meaning (e.g., blue skies)

16. Applying inconsistent decision rule across different areas of analysis
17. Considering data as proof rather than as probability
18. Reducing consideration of causality to a few variables and failing to consider other potential causes
19. Assuming causality or a pattern for random events (apophenia)
20. Assuming an either-or duality rather than an intermediary position between the extremes (law of an excluded middle)
21. Ignoring principle of contradiction whereby a conclusion or statement cannot be simultaneously true and false
22. Not indicating loss or cost of an erroneous decision
23. Applying overconfidence in minimizing risk and overestimating chance of success

24. Extending results to a ridiculous degree to invalidate an argument (*reductio ad absurdum*)
25. Embarking on the "slippery slope" of extending results or conclusion to a ridiculous degree to validate an argument
26. Displaying cognitive dissonance in justifying decisions by blaming others or situation when confronted with contradictory information
27. Recalling past beliefs or behaviors as consistent with current beliefs or behaviors
28. Assigning favorable views to members of your group and unfavorable views to other groups
29. Applying attribution bias whereby adverse results are caused by external factors in the situation or behavior of others and positive results are caused by personal effort
30. Applying attribution bias whereby personal beliefs are rational and beliefs of others are emotional or irrational
31. Self-fulfilling through behaving or believing in ways that conform to expectations
32. Attributing human characteristics to inanimate or ambiguous objects (personification)
33. Distorting memory such that false information is retained longer than accurate information
34. Forming different conclusions based upon a limited number of possible options provided (framing effect)
35. Making no choice when confronted with too many options
36. Indulging an aversion to extremes and maintaining status quo or intermediary choice in a range
37. Indulging loss aversion in belief that possible loss is more important than possible gain
38. Accepting a value as the standard of comparison to other options (anchoring)
39. Valuing what we own or do more than what we do not own or do (endowment effect)
40. Applying Weber's law (absolute increase matters less than proportional increase)
41. Allowing time pressure and/or emotional persuasiveness to influence behaviors or beliefs
42. Allowing "sunk cost" of past cost or effort to affect future decisions
43. Believing that attractive persons have greater creditability
44. Applying political bias to create false balance for those who have already decided
45. Allowing decision to be affected by perceived rarity or uniqueness

EXAMPLES

- President Bush often paraphrased the saying, "You're with us or against us." This is an error of the law an excluded middle. Any intermediate or alternative positions, such as trying a different tactic, are not included.
- Jerome Groopman, in *How Doctors Think*, states that doctors misdiagnose up to 40% of conditions, mostly due to the following three cognitive errors:

    1. Anchoring in which a judgment is based on the first or anchor symptom
    2. Using availability of easily remembered prior experience to explain or make the diagnosis
    3. Basing the symptoms used to diagnose upon the stereotyped attributes of the patient

- *The Plain Dealer* reported that a Firestone Centennial Driving Survey conducted by Bridgestone/Firestone, Inc., found that only 29% of men ask for directions when they have made an error and are lost, but 64% of women ask for directions when they have made an error and are lost. This is an error of overconfidence more likely in males. (The study did not report which gender more often made hand gestures when corrected for a driving mistake.)
- Kevin O'Brien, deputy editorial page director of *The Plain Dealer*, wrote that Obamacare was for "people who are too stupid, incompetent and weak to manage their own affairs." O'Brien also stated that the proposed legislation was a health care takeover bill that had "socialistic roots." (Do you think there are any personal biases in this opinion? How about this illogical statement: "Repeal the government health takeover bill, but don't take away my Medicare"?)
- CNNPolitics.com quoted Tom Coburn, Republican senator from Oklahoma, as saying, "So don't catch yourself being biased by Fox News that somebody is no good." (Does the media influence perhaps contribute to erroneous interpretations of people and events?)
- Daniel Kahneman states, in *Thinking, Fast and Slow*, that 65% of small businesses fail in the first five years. The probability that a person will find success is only approximately one in three. Is this a significant difficulty? Are we sure that we will succeed? What is the operational definition of a small business? With odds like these, why do so many think that they can be successful? Is this an error of overconfidence?
- The delayed discovery of the cause of the disease Pellagra is a historical example from the early 1900s of the consequences of errors resulting from personal bias in interpreting information according to current beliefs. The

prevailing scientific assumption at that time was that this disease, like so many others, was infectious and caused by a germ. Ineffective treatments for cleanliness contributed to thousands of deaths. Objective analysis indicated that the disease was more prevalent among poorer individuals who had a low-protein diet. Pellagra is caused by vitamin deficiency, which can be prevented by an appropriately balanced diet. This example indicates the benefit of changing beliefs to accommodate existing facts rather than trying to make the facts fit into existing beliefs. Our personal biases are not just a point of discussion; they have significant consequences for the decisions made by individuals or as a society and can literally mean the difference between life and death.

REDUCING ERRORS

Reducing errors due to our biases requires both a willingness to acknowledge our personal limitations and a commitment to implement appropriate research procedures to gather and interpret information. Courage to confront our biases and acknowledge our limitations may be a prerequisite for intellectual growth.

Understanding our own biases may be a first step in reducing errors. This chapter presented numerous types of biases and cognitive errors. Each of us must become aware and adopt the discipline of seeking to identify our biases to better understand and interpret information. The disposition to modify our beliefs and behaviors when more accurate information becomes available is a recurring theme in the effort to reduce our biases and erroneous decisions.

A second step is to use the multiple types of errors identified in this textbook as a checklist for carefully examining the accuracy of the information. Caution is required in interpretation whenever errors occur in any of the research processes.

Whatever the error, the third step for researchers or consumers of information is to understand how it affects the accuracy of the information and the interpretation. Specific types of errors may have significant adverse consequences for making accurate decisions. Some errors are minimal and simply limit the range of responses or the generalizability of the results. A modified interpretation that provides an accurate, but more limited, conclusion reduces the adverse effects of these errors.

It is more difficult, if not impossible, to modify the adverse effect on the accuracy of interpretations or decisions of significant errors involving inappropriate design, invalid or unreliable measurement instruments, or ethical errors involving fabrication or falsification of data. In these situations, conclusions may not be possible.

BEST PRACTICES

- Acknowledge that the research results are tentative rather than absolute.
- Use a checklist to identify potential errors described in this text.
- Use a checklist to identify alternative explanations for results described in this text.
- Use a checklist to identify best practices in experimental, statistical, and other controls described in this text to minimize errors and alternative explanations.
- Examine the list of common cognitive and interpretation errors described in this chapter to identify and minimize these errors in decision making.

INTERRELATION TO OTHER TOPICS

This chapter showed that each of us has many biases and cognitive errors in processing information. The use of checklists to identify errors can provide some protection from errors caused by our biases. The adoption of a disposition to modify beliefs and behaviors when accurate information is available can also help to reduce adverse effects from these errors.

The next chapter presents a topic that is disturbing to the functioning of a society and, potentially, to the well-being of each of us. Individuals, companies, and agencies will sometimes intentionally create and communicate inaccurate information. Identifying erroneous information may help to reduce the influence of misinformation.

Chapter Thirteen

The Age of Misinformation

"The Gripes of Wrath"

Enlightten the people generally, and tyranny and repression of body and mind will vanish like evil spirits at the dawn of day.

—Thomas Jefferson

Ignorance is preferable to error; and he is less remote from the truth who believes nothing, than he who believes what is wrong.

—Thomas Jefferson

He who permits himself to tell a lie once, finds it much easier to do it a second and third time, till at length it becomes habitual.

—Thomas Jefferson

Facts are stubborn things; and whatever may be our wishes, our inclinations, or the dictates of our passions, they cannot alter the state of facts and evidence.

—John Adams

Thou shall not bear false witness against thy neighbor.

—The Holy Bible

Put off falsehood and speak truthfully.

—The Holy Bible

Never give a sucker an even break.

—Edward Albee

Time's glory is to calm contending kings, to unmask falsehood and bring truth to light.

—Shakespeare

There is only one good, knowledge, and one evil, ignorance.

—Socrates

To first understand the lies, you first have to understand the truth.

—Paul Krugman

SIGNIFICANCE

We live in a world of so much available information. The airwaves and media reports are congested with conflicting views that make it difficult to penetrate the fog of misinformation. We continue to look for the figurative lighthouses or listen for foghorns to guide us in avoiding erroneous information.

Individuals and groups purposively create misinformation through distortion or manipulation to influence the beliefs and behaviors of others. The goals of the misinformation are often money, political power, the spread of a religious belief, or interpersonal power. Misinformation is often the only avenue to accomplish these goals. The blunt realization is that these goals are unattainable with accurate information.

Separating errors from accurate information is an important task for individuals and society. The disturbing realization is that few sources are available to identify errors and misinformation. Society tolerates a mild degree of misinformation. Real estate agents may engage in "puffery," using statements in which they inflate or enhance the quality of a property that they are selling. The common phrase "caveat emptor," or buyer beware, is used to indicate that the buyer or consumer of a product has responsibility for determining the accuracy of information in making decisions. Do people wear makeup to present an enhanced appearance?

The issue is not with these widely tolerated examples of misinformation or the numerous and often humorous billboards that advertise something as "world famous" or the misleading comments posted to online dating sites. The serious concern is with the creation and dissemination of misinformation that has profound adverse effects, resulting in losses of money, life, and personal freedoms.

Misinformation is found in all types of communication methods, and for almost every topic. This textbook has provided hundreds of types of errors and representative examples of errors in different research procedures. Intentional misinformation reveals a dark and troubling side of humanity.

LEARNING OBJECTIVES

1. Understand that some information is misinformation, intentionally inaccurately collected, analyzed, interpreted, and reported to manipulate the opinions and decisions of others.
2. Understand that goals of the misinformation can be monetary, political, religious, or interpersonal.
3. Understand that it is necessary to examine all information or data obtained for any errors.
4. Understand that attempts to suppress research also contribute to misinformation.
5. Understand that cognitive bias and errors affect how information is processed.

PREREQUISITES

1. Understand that beliefs change when facts change.
2. Understand the various errors in research data that limit the accuracy of results.
3. Examine all information for factual accuracy.

CONCEPTS

All of us have experienced an uncomfortable situation in which we made an incorrect decision because we were misled or deceived or because facts had been withheld. We may have suffered in silence or become angry. We may have vowed to never again be misled. Sadly, people, groups, or organizations purposely try to deceive us for monetary, political, religious, or interpersonal gain.

The world is bursting with information. For example, the University of California, Berkeley has indicated that as many as 1.1 trillion e-mail messages are sent each year. We also see expansion in all forms of information, including research information. At what point does all of the information become so overwhelming that it becomes confusing? What may be required is not just more data. We also need more accurate data that are free from error, more knowledge in understanding errors, and more wisdom in accurately interpreting the information.

The highest bidder purchases our personal data. Privacy is almost obsolete, and this places each of us at risk of increasingly targeted ads and, unfortunately, misinformation. Many organizations reflect BIG DATA in that they know where we live, our preferences, and what type of message will be most effective in manipulating our beliefs and behaviors. In this situation, the danger of all of the available information is not utopia, but a dystopia created by misinformation.

This textbook has provided us with an understanding of the many types of errors and inaccuracies in the information around us. A dominant purpose of this text is to help us to be better consumers of information. These errors and inaccuracies are not just a figment of our imagination; they are real and can adversely affect our decisions in many areas of our lives. The hope is that we choose wisely.

COMMON ERRORS

1. Using all of the errors noted in previous chapters of this text
2. Attempting to suppress, distort, or manipulate research information

EXAMPLES

It is impossible to provide examples of all of the types of misinformation that we face each day, through blogs, mass e-mails, talk radio, cable TV, Internet, op-ed pages of newspapers, commercial advertisements, political advertisements, conversations with our friends, and so many other sources. It is almost as though our personal security threat level for inaccurate information has moved from orange to red.

The challenge that we all face is to distinguish between the accurate and inaccurate information. Faith in the use of reason to have logical debates using the best available information is a foundation of American democracy. We all have a responsibility to understand and identify erroneous information in order to make the best decisions.

Why do we keep falling for inaccurate information? As indicated in a previous chapter, we are predisposed to superficially process information according to our biases and lack of understanding of research methods.

Among the most prominent of these biases are seeking information that confirms our beliefs and ignoring information that disconfirms our beliefs; basing decisions upon easily available information; basing decisions upon vivid information; using emotionally charged labeling rather than more neutral language; making statements that inaccurately or imprecisely describe; and following group pressure to agree with a belief.

Our inability or unwillingness to examine the information available to us represents a victory for misinformation. False knowledge is a significant problem in our society. Various media cover and perpetuate "myth information" to gain listeners and market share and are very real forces that can adversely affect everyone in our society.

There are not 50 shades of the truth. In all of the information that we use, we should strive as a society for "the truth, the whole truth, and nothing but the truth."

- *The New York Times* reported that the Union of Concerned Scientists, which comprises over 12,000 scientists, including 52 Nobel laureates, urged the Bush administration to stop suppressing reports and then including erroneous information in scientific documents. The affected reports covered topics ranging from drug safety to childhood lead poisoning, mercury and environmental pollution, and climate research. The scientists signed this statement to urge the end of the distortion of scientific knowledge and the restoration of scientific integrity as vital to the well-being of a democracy.

  Over 50 Nobel laureates demonstrated the courage to protest against the errors of misinformation. Do we have that courage?
- David Barstow, reporter for *The New York Times*, sued the Department of Defense for documents regarding the relationship between Pentagon officials and former military officers who presented themselves as independent observers on television. The former military officers did not reveal that they had been coached by the Pentagon, had profited from their relationships with military contractors, and had been threatened with loss of power through loss of access to classified information if their comments did not reflect Pentagon talking points. (See any ethical errors regarding potential conflict of interest.)
- *USA Today* reported that the National Rifle Association has used some of the same tactics employed by the tobacco lobby to block gun-control laws. These tactics include blocking research, suppressing information, and discrediting science. A 17-year ban was in effect to block research that could be used to "advocate or promote gun control." Could suppressing research create errors of misinformation?
- As reported by *USA Today* and many other news services, Johnson & Johnson paid more than $2.2 billion in a health fraud settlement for criminal and civil claims that it marketed Risperdal and other drugs for unapproved uses. The company ignored government warnings and failed to publish studies that confirmed the risks of the drug. The company told its sales force to market the drug as a treatment for children despite increased health risks.
- *The New York Times* reported that a series of large settlements have been reached in recent years involving allegations of improper marketing by pharmaceutical companies, including GlaxoSmithKline ($3 billion) and Pfizer ($2.3 billion). David Stone, attorney, commented that incentives remain for these companies to make money by disregarding Food and Drug Administration regulations.

- A classic case of misinformation is the Piltdown Man hoax, in which human and nonhuman bone fragments were claimed to have been excavated as part of the same specimen. The claim was that this specimen was the missing link between apes and humans that supported the theory of human evolution. This claim confirmed existing beliefs about evolution as held for decades. More advanced dating techniques exposed this as a hoax.
- A common and often repeated misinformation technique is the use of "average." A statistical average is not what most individuals will receive or do. Numerous examples of this error are present in everyday life. A few extreme scores can distort the average.

REDUCING ERRORS

There are many more examples of disseminating misinformation from the past and the present. Misinformation can be appealing if it confirms current beliefs or can change the behaviors of others to accomplish financial, political, religious, or interpersonal goals. Until human nature changes, misinformation will continue to be disseminated.

Reducing or preventing misinformation may not be possible. What is possible is to gain knowledge about identifying errors in gathering and interpreting information.

Identification of errors is necessary, but not sufficient, to reduce the possibility of using misinformation. A disposition to change beliefs and behaviors—based upon factually verifiable information—is also required. It does no good to know the error of our ways if we do not also possess the courage to change.

Misinformation has often contributed to heated discussions and conflicts that have polluted the opportunities for sharing and cooperating. We live in an age of distraction in which we do not always carefully examine the accuracy of the information that we receive. A checklist can assist in identifying those errors.

BEST PRACTICES

- Examine multiple sources of information.
- Use a checklist of the errors described in this textbook to examine each source of information for potential errors.
- Change beliefs and behaviors as additional factual information is obtained.

INTERRELATION TO OTHER TOPICS

We have examined multiple research procedures in this textbook to understand potential errors and enhance our ability to make accurate decisions. We need to become knowledgeable, but also have the disposition to seek truth and change our beliefs as the facts change.

The next chapter presents the hope that religion and science can provide opportunities for understanding the world around us.

Chapter Fourteen

Religion and Science

"Search for the Fountain of Truth"

Faith and reason are like two wings on which the human spirit rises to the contemplation of truth.

—Pope John Paul II

Jesus of Nazareth was the most scientific man that ever trod the globe. He plunged beneath the material surface of things, and found the spiritual cause.

—Mary Baker Eddy

Scientific thought is not an accompaniment or condition of human progress, but human progress itself.

—William Clifford

If we do discover a complete [unified] theory [of the universe], . . . it would be the ultimate triumph of human reason—for then we should know the mind of God.

—Stephen Hawking

Over the entrance to the gates of the temple of science are written the words: Ye must have faith. It is a quality which the scientist cannot dispense with.

—Max Planck

Faith which does not doubt is dead faith.

—Miguel de Unamuno

Politics, like religion, hold up the torches of martyrdom to the reformers of error.

—Thomas Jefferson

I myself believe that the evidence for God lies primarily in inner personal experiences.

—William James

The God that science recognizes must be a God of universal laws.

—William James

Science deals mainly with facts; religion deals mainly with values. The two are not rivals.

—Martin Luther King

Accurate information is not the enemy of informed faith.

—Pastor David Collings

My concern is not whether God is on our side. . . . My concern is whether we are on God's Side.

—Abraham Lincoln

SIGNIFICANCE

The dialogue between religion and science is not just about differences of academic opinion. The varying opinions provide a base for many significant actual personal and societal decisions. Many issues are involved in examining religion and science, including human emotions, different forms of knowledge, the nature of evidence, randomness versus purpose, and certainty versus tentativeness of knowledge. These issues have significant influence in the religion-science dialogue.

As in any beneficial dialogue, each side may learn from the other. The scientific goals of objectivity and reason must also incorporate the powerful religious goals of faith, purpose, and good works. Both sides of the dialogue need to be included to gain a more complete knowledge of the world around us.

Religion and science should also be open to continuing to gain knowledge and modifying our understanding of the world as more facts becomes available. Developing a common knowledge of how people process information and behave can make both religion and science recognize the limitations to their view and acknowledge the contributions of the other.

Textbooks about research procedures include little information regarding dispositions, cognitive biases, and religious influences. These topics are included in this text because of their significant effects on how research information is gathered and interpreted. Understanding errors in research requires that we examine sources of errors, even if the errors reflect our dispositions, biases, and beliefs.

Examples are included in this chapter of errors in the interpretation of religious and scientific evidence. Some of these errors have contributed to significant adverse effects and even loss of life.

LEARNING OBJECTIVES

1. Understand that science and religion use different methods to gather and interpret data.
2. Understand that both scientific practices and religious practices have changed over time.

PREREQUISITES

1. Be able to change beliefs when facts change.
2. Understand the various errors in research data that limit the accuracy of results.
3. Understand the concepts of a religion and the inconsistencies that may be present.

CONCEPTS

No topic involving research has been the focus of more energetic debate and discussion than the relative merits of using religion or science to understand and predict events. Why is there such an apparent confrontation? The answer appears to reflect a combination of the different cognitive biases and errors that we all have and our disposition regarding anxiety in dealing with uncertainty and new information.

We do not see a universal attitude of religion toward science. The Institute for Studies of Religion at Baylor University, for example, identified that Americans have four different and major images of God: as authoritarian, benevolent, critical, and distant. These images are a powerful predictor of social and political views. The acceptance of science varies according to which of the four images of God a person holds. The inference is, then, that attitudes toward science and religion are part of a broader personality of the individual. Some people are disposed to accept the certainties promised by many religions and feel threatened by the uncertainty that might be an inevitable part of scientific investigation. Persons with this disposition would rather accept the literal meaning of religious writings.

Is allowing a literal reading of the Bible to guide our decisions an error? During the Civil War, many religious leaders in the South used the literal meaning of the Bible to defend slavery, quoting from Titus 2:9: "Tell slaves to be submissive to their masters and to give satisfaction in every respect." In addition, Genesis 1 tells people to "be fruitful and multiply." Does this condemn relationships that do not produce children?

Aramaic was the language generally spoken in the areas around Jerusalem at the time of Jesus. How accurate are the translations into Greek or English that we now use to determine the literal meaning of the Bible? Could translation errors exist? What are the implications if we find translation errors?

The cognitive and dispositional biases and errors included in previous chapters may also have significant implications for how we view religion and science. The most significant of these errors include confirmation bias, cognitive dissonance, social pressure, dislike of uncertainty, and simplification by considering only a few rules or causes.

The most significant dispositional errors appear to be reluctance to admit that we made a mistake and unwillingness to change our beliefs when the facts change. This combination of cognitive and dispositional errors is a poisonous concoction for humanity.

Does science explain religion? Does religion explain science? Do religion and science enrich the explanations of each other? The answer appears to be that religion and science enrich each other, as eloquently stated by Pope John Paul II in the Jubilee of Scientists on May 25, 2000: "It can be said, in fact, that research . . . contributes to the glory of God which is reflected in every part of the universe."

Every scientific investigation requires ethical considerations. A commitment to avoiding errors in gathering and interpreting information is an ethical and moral issue. Are scientists true to the disposition that there may still be unknown causes for events? Do religious leaders accept that science continues to discover new information about causes of events? Do scientists and religious leaders accept the confluence of the idea that research can contribute to the glory of God?

The evidence that would answer all of our questions about nature and religion is present and awaits our discovery. Multiple methods of investigation will provide a more complete understanding. Here is an important question: Do our beliefs determine the facts or do the facts determine our beliefs?

COMMON ERRORS

1. Believing that scientists cannot have religious faith and religious persons cannot have faith in scientific research
2. Failing to change beliefs when new facts become available

3. Seeing only what we want to believe
4. Believing that religious scripture represents absolute truth
5. Believing that science discovers absolute truth

EXAMPLES

- Public Policy Polling found that even in 2013, 26% of Americans believed or were not sure that President Barack Obama is the anti-Christ. Does this belief affect perceptions of Obama's policies?
- Pope John Paul II stated in 1996 that evolutionary theory is compatible with Catholic teaching, but Cardinal Christopher Schönborn stated in *The New York Times* that the pope's statement was "vague and unimportant." Peter Raven, a member of the Pontifical Academy of Sciences, stated in *Science* that Cardinal Schönborn had a "serious misunderstanding" of evolution and the view of the Church. It is an error to assume that religious persons cannot have faith in scientific research.
- CNN.com reported that the National Association of Evangelicals (NAE) had rejected a letter signed by prominent religious conservatives to prohibit the NAE policy director from urging action on global warming. One of the signers of the letter, Tony Perkins of the Family Research Council, stated that global warming is part of a leftist agenda to divide evangelical unity. A member of the NAE board said, "Environmental degradation is an offense against God."

  (What is the relationship between religion and science? Why do religious persons differ on the roles of religion and science? Is a religious view an extension of a broader personal or political position?)
- *The Plain Dealer* reported that in 1992 Pope John Paul II declared the Church's 17th-century denunciation of Galileo's idea that the earth revolved around the sun to be an error. (The Pew Research Center reports that 18% of Americans still believe the sun goes around the earth!)
- CNNPolitics.com reported that the United States Conference of Catholic Bishops had criticized the Patient Protection and Affordable Care Act as not being restrictive enough on abortion and made an erroneous claim that the bill will provide taxpayer funding for elective abortion. The Catholic Health Association of the United States and a group of Catholic nuns urged passage of the bill. According to the Guttmacher Institute, Catholic women account for 31.3% of the approximately 3,700 abortions performed each day in the United States. (Is it an error for one group to speak for all in a religion? Can different groups within a religion have different views? Is religion a personal matter or do all followers need to obey the decisions of a particular group with the religion?)

- The Gallup Poll reported that 46% of Americans believe that God created human beings very similar to our present form within the past 10,000 years. (Could God have created human beings in a less similar form millions of years ago? Why are people frightened by the possibility that humans may have changed over the years? Is it a fear that humans may further change in the future?)

REDUCING ERRORS

Truth is a goal shared by religion and science. Avoiding errors is a process that religion and science can use to understand the world. Religion and science must each appreciate the methods used by the other to gather and interpret information. For religion, this means the ability to change beliefs as more research becomes available about the physiological and cognitive processes that humans use to perceive and process information. For science, this means appreciation for the human need to perceive the possibility of a higher power and a purpose for life.

The goal of true understanding is compatible with both religion and science. The methods of inquiry used by religion and science are complementary, with each providing different facets of a more comprehensive understanding. The important dispositions for each are to accept the validity of the methods of the other and to change beliefs or understanding as more facts become available. The pursuit of truth and a willingness to change helps to reduce errors.

Research is compatible with both religion and science. Religion and science share the optimism of discovering a greater understanding.

BEST PRACTICES

- Accept that religion and science seek truth.
- Accept that religion and science may learn from each other.
- Be humble and acknowledge the uncertainty of current knowledge of the world around us.

INTERRELATION TO OTHER TOPICS

The relationship between religion and science has been long and often antagonistic. Despite this antagonism, religion and science have a common ground in seeking the truth for the betterment of humanity. Avoiding errors helps each to gain a more accurate understanding.

The epilogue of this textbook gives an example of how research can be redesigned to correct errors in previous research. Knowledge of potential errors and application of the best practices in this textbook will help, but we also need courage and the disposition to change as the facts change.

Epilogue

How Will We Apply Research to Improve Our Lives?

The great use of life is to spend it for something that will outlast it.

—William James

The spirit of truth and the spirit of freedom—they are the pillars of society.

—Henrik Ibsen

My holy of holies . . . and the most absolute freedom imaginable, freedom from violence and lies, no matter what form the latter two take.

—Anton Chekhov

The truth shall make you free.

—The Holy Bible

The condition upon which God hath given liberty to man is eternal vigilance.

—John Curran

I shall try to correct errors when shown to be errors; and I shall adopt new views so fast as they shall appear to be true views. . . . I intend no modification of my oft expressed personal wish that all men, everywhere, could be free.

—Abraham Lincoln

We should all be concerned about the future because we will have to spend the rest of our lives there.

—Charles Kettering

We are not afraid to follow truth wherever it may lead, nor to tolerate any error so long as reason is left free to combat it.

—Thomas Jefferson

To err is human; to identify and correct errors shows humanity.

—Anonymous (OK, it's us again)

If I have seen further it is by standing on the shoulders of giants.

—Sir Isaac Newton

I do not know what I may appear to the world; but to myself I seem to have been only like a boy on the seashore . . . whilst the great ocean of truth lay all undiscovered before me.

—Sir Isaac Newton

Research is an integral and dynamic part of human life, from our earliest ancestors to the present. Research literacy in understanding and applying the procedures to identify accurate and erroneous information can help in all decisions that we make to accomplish our goals.

Unrecognized and/or uncorrected errors, based upon inaccurate or incomplete (weak) application of research procedures, can contribute to incorrect beliefs and decisions. This textbook includes hundreds of possible errors in conducting research and some suggestions for best practices to avoid and/or correct these errors.

The quality of research is dependent upon the accuracy of all of these procedures. Errors involving any single research procedure can invalidate the entire research. For example, a highly valid and reliable data collection instrument cannot compensate for an unrepresentative sample.

Research is a continual process of correcting errors or weak application of procedures by redesigning studies to test hypotheses regarding other potential explanations. If the alternative hypotheses are not supported, then we can "redesign" the study, testing new variables, with new methods in pursuit of knowledge and improving the world. Like the ancient city of Troy, which has multiple cities built upon the foundations of previous cities, knowledge builds upon previous research, and this process can continue over centuries with different researchers and procedures.

An example of this continual redesign process examining other potential explanations is a study by Richard Wiesenthal, Bruce Cooper, Sheldon Marcus, and Ruth Greenblatt, published in a 1997 issue of the *Journal of Educational Administration*. This study used the Homework Attitude and Behavior Inventory for Teachers (HABIT) to obtain information regarding teacher beliefs about homework and school. Teachers in schools with a well-defined homework policy gave more homework than teachers in schools without a well-defined homework policy.

Redesign of the study with additional independent, extraneous, and dependent variables can make the study more accurate and sensitive. Do the *subject* (e.g., math, science, music, English) taught and the *age and grade level* (e.g., elementary versus high school kids) affect the number of times homework is assigned? Does the *time of year* (e.g., early, middle, or late in the semester), the *day of the week* (Monday versus Friday), or the *type of homework* assigned matter? Does *parental attitude* toward homework affect student performance?

A redesign of research with testable hypotheses examined these other potential explanations. The HABIT questionnaire responses and weekly homework behaviors of teachers comprised the data. As you can see, research is a continual process of redesigning and testing hypotheses regarding other potential explanations. See the framework and variables below for a redesign of the study with additional variables to obtain greater understanding:

Table E.1. Research Design

Independent Variables	Extraneous, Mediating, and Predictor Variables	Dependent Variables
• Focused versus unfocused homework policies **Descriptive Categories** • Gender • Subject taught • Highest academic degree	**HABIT Attitudes and Beliefs** • Student work habits • Student knowledge acquisition • Student attitudes • Teacher-centered attitudes • Parental involvement • School climate	**Homework Behaviors** • How often homework is assigned • How often collected • How often graded • How often returned **Types of Homework** • Repeating classwork • Teaching new skills • Introducing new topics • Imaginative pursuits

The focus of this introductory textbook is on the identification of errors in research. Many real-world examples are included. The great variety of research designs used to avoid and/or correct errors to increase understanding is beyond the scope of this introductory textbook. A companion text will provide real-world examples of alternative research redesign methods that supply accurate information.

The following prescriptive advice can help each of us to redesign our lives to avoid and/or correct errors in our beliefs or decisions:

RESEARCH PRESCRIPTIONS FOR AVOIDING AND/OR CORRECTING ERRORS

- Accept errors as inaccurate beliefs or decisions that result from inaccurate or incomplete (weak) research procedures or interpretation of research results (chapter 1).
- Accept that research provides only tentative conclusions because of unrecognized or uncorrected errors (chapter 1).
- Change beliefs as facts change (chapter 2).
- Conduct research in an ethical manner (chapter 2).
- Use operational definitions to develop testable hypotheses (chapters 3 and 4).
- Use data collection instruments with adequate validity and reliability (chapters 5 and 6).
- Select samples that accurately represent the intended population (chapter 7).
- Select and implement research designs that control or reduce the threats to the validity of research (chapters 8 and 9).
- Select methods of data analysis that are appropriate for the characteristics of the obtained data (chapter 10).
- Interpret results in terms of probability (chapter 11) rather than with personal and cognitive biases (chapter 12).

To paraphrase Newton's quote at the beginning of this chapter, the hope is that each of us can now go beyond standing at the shore of the great ocean of truth. These research prescriptions can serve as a beacon, helping us to avoid the shoals of errors as we calmly sail in that great ocean of discovery.

The most challenging task that we face may be to acquire and develop the disposition to openly seek accurate information and be willing to change our understanding as the facts change. We may use our mind to identify errors, but changing our beliefs when more facts are available requires changing both the heart and the mind.

It is not too late to become the person that we always wanted to be. If we do not like our current situation as an ending, then it is not the end. We cannot prevent the past, but we can create a new ending. The hope is that we will use research to help create a future that we want and that is free from error.

Continue to learn humbly, as though you will live forever. Choose the paths that benefit others and make the earth a better place than when you arrived.

Index

www.ingramcontent.com/pod-product-compliance
Ingram Content Group UK Ltd.
Pitfield, Milton Keynes, MK11 3LW, UK
UKHW012040150926
475560UK00001B/105